SpringerBriefs in Applied Sciences and Technology

For further volumes:
http://www.springer.com/series/8884

Amin Ranj Bar • Muthucumaru Maheswaran

Confidentiality and Integrity in Crowdsourcing Systems

 Springer

Amin Ranj Bar
McGill University
Montreal
Québec
Canada

Muthucumaru Maheswaran
McGill University
Montreal
Québec
Canada

ISSN 2191-530X   ISSN 2191-5318 (electronic)
ISBN 978-3-319-02716-6   ISBN 978-3-319-02717-3 (eBook)
DOI 10.1007/978-3-319-02717-3
Springer Cham Heidelberg New York Dordrecht London

Library of Congress Control Number: 2013951524

Printed on acid-free paper

Springer is part of Springer Science+Business Media (www.springer.com)

Contents

Chapter 1
Introduction

Crowdsourcing information systems enlist a crowd of users to collaboratively build wide variety of information artifacts. Over the past decade, a large number of crowdsourcing information systems have been developed on the Internet such as Wikipedia, Linux, Yahoo! Answers, Stack Exchange, and numerous online social networks (OSNs). Because the development of crowdsourcing information systems is an emerging area, it has appeared under many names, including peer production [190], user-powered systems [80], user-generated content [80], collaborative systems [109], community systems [77], collective intelligence [214], wikinomics [199], crowd wisdom [195], smart mobs [171], mass collaboration [172], and human computation [80].

There are several motivating factors for using crowdsourcing. Chief among them is the ability to leverage the wisdom of the crowds. Many problems including protein folding, website design, developing advertising campaigns, and image labeling have benefited from the wisdom of the crowds. The "wisdom" of the crowd means different things for different problems. For instance in protein folding and website designs, crowd wisdom brings diverse heuristic ideas to solve the problem, which could lead to faster or higher quality solutions. In the image labeling problem the crowd could use superior image recognition skills of the humans to easily classify the images [80]. The crowd could bring a very different type of benefit for creating advertising campaigns. Because the crowd is created from the target consumer population, the ideas brought into the campaign by the crowd is bound to resonate well with the eventual audience of the advertisements. Another motivating factor for crowdsourcing is the cheap and anytime accessible labor. With the massive connectivity created by the Internet, with clear participatory benefits, organizing large crowds is feasible. People can participate in carrying out the crowd task in their free time.

Despite the compelling benefits of crowdsourcing, there are many challenges that need overcoming to leverage the full potential of crowdsourcing. As the novelty of crowdsourcing has already worn out, crowdsourcing efforts require clear incentives to gain the interest of large number of people. So far, crowdsourcing systems have used different forms of incentives such as: community recognition (e.g., Stackoverflow, Linux), monetary gain (e.g., Amazon Mechanical Turk), game experience (e.g., online ESP Game), and community goodwill (e.g., Wikipedia). One of the challenges

A. Ranj Bar, M. Maheswaran, *Confidentiality and Integrity in Crowdsourcing Systems,* SpringerBriefs in Applied Sciences and Technology, DOI 10.1007/978-3-319-02717-3_1, © The Author(s) 2014

facing crowdsourcing is the identity of the participants. Because asserting the identity on the Internet remains an unsolved problem, it is impossible to determine whether a new user joining the crowd is actually a miscreant evicted in the past from the system. Such recurrent miscreants can reduce the efficiency of a crowdsourcing system by increasing the workload for the productive users.

Another challenge is information privacy in crowdsourcing information systems. Controlling the confidentiality of the information generated by the users is a major concern here. Crowdsourcing information systems can be categorized into two classes based on who owns the user-generated content: (i) creating user and (ii) collective community. Online social networks (OSNs) is an example of crowdsourcing systems, where the content is owned by the creating users. The OSNs can be categorized into three major categories: friendship networks, common interest networks and interaction networks. In friendship networks, people create profile pages that describe themselves and that explicitly link them to their friends' profiles. There are many OSNs that fall into this category such as Facebook, MySpace, and LinkedIn. In common interest networks, social networks emerge as users collaboratively create some online content such as photo albums, wikis, and blogs. In interaction networks, social networks are defined by the communication patterns used in instant messaging services or email. OSNs are dynamic networks with topological changes that are caused by edge and node creations or deletions. With the recent popularity of OSNs (Facebook counts more than one billion users), the information sharing activity in these networks is emerging as an important privacy issue.

As OSNs increase in size and more people use them as their primary Internet portal, the volume of information shared in OSNs keeps on growing. Information is created by different sources in OSNs including people posting information in their profile pages, relational information generated by people initiating connections among themselves, and data feeds generated by sensing people's activities such as gaming and purchasing. OSNs store and process different pieces of information that are related to users: picture files, relationships among people, and sharing preferences regarding data objects. This means the security primitives built into the OSNs play key roles in defining information security in the social web. Consequently, many research thrusts have examined wide-ranging security issues in the context of OSNs [57, 204, 39, 65, 89, 88, 90, 187].

While information sharing is vital for socializing online, it is necessary to ensure users that their privacy and access control requirements are preserved when information sharing occurs within OSNs. Recently, users in OSNs have started to become more aware of the risk of unauthorized propagation of their information through social networking sites. The recent complaints received by Facebook for the "free Facebook t-shirt" [144] or "Facebook phonebook contacts" [161] are examples of information exposure in OSNs. To partially address users' needs to control access to their data, several topology-based access control mechanisms for OSNs were proposed to define users who are authorized to have access to data objects by specifying constraints on the social graph [62, 39, 37, 81, 85, 125].

Because existing techniques only deal with information release, a user might not be able to precisely identify who is authorized to have access to her data. Even in small social networks, it is difficult for a user to keep track of the topology of her constantly changing social network and to identify users who are actually authorized even with simple access rules such as "friends-of-friends." In addition, the user's privacy requirements are constantly changing [41, 36, 42]. Users can lose control of their shared data and risks of unauthorized dissemination of their data escalates with increasing number of social interactions [63]. Specially, a user may not be able to track how her private information is handled by her friends after she has released the information to them [63]. The topology based access control mechanisms give a static control scheme based on particular friendship configurations. Therefore, it is necessary to have new access control mechanisms in OSNs to evaluate the potential risks and to make users fully aware of the possible consequences of their decisions in specifying access rules.

Yet another challenge is the integrity of the information generated in crowdsourcing systems. The major problem with integrity management is defining integrity itself. In computer security [49], data is considered to have high integrity if it is not tampered by unauthorized modifications. With crowdsourcing, it is not possible to group the modifications as authorized and unauthorized. Therefore, popular crowdsourcing systems such as Wikipedia that want to generate high quality content rely on consensus [71] among the users. The consensus could be arrived using a single arbiter (e.g., the Linux model) or using distributed collection of editors (e.g., the Wikipedia model) [189, 197].

To meet its integrity objectives, Wikipedia encourages contributions by making it easy for the contributors to create and update any article. The integrity of the contributions are checked and flagged by subsequent readers. For highly trafficked articles, this model of integrity enforcement works very well. In Linux, integrity is given very high priority. All updates submitted by the development community need the final approval of the project originator (Linus Torvalds) before they are included in the official software release. Community feedback and importance of the contribution are some of the factors that can influence Linus Torvalds' decision to include or exclude the contribution.

Several Wikipedia-like projects that do not have the popularity of Wikipedia use a model similar to that of Linux's to manage the integrity of the articles maintained by them. However, instead of relying on a single person for the whole project, these sites [14, 20] designate the article creator as the person responsible for the integrity of the article. While having a central figure per article facilitates integrity maintenance, it can create lots of workload for the maintainer if there are frequent updates from the community.

Mainly, there are three problems to preserve integrity of articles on online crowdsourcing systems. The leading problem is the lack of authority in large-scale collaborative content sharing websites such as Wikipedia [216]. For instance, readers of Wikipedia cannot know who has written or modified the article they are reading, it may or may not have been written by an expert. The second problem is the lack of content verification on specialized topics. Someone should report the problem;

otherwise, inaccurate information that is not obviously false may exist in Wikipedia for a long time before it is challenged [1, 224]. Lack of fact checking may result in biased articles with various contentions that will need further resolutions. As ways of addressing these problems crowdsourcing systems allow multiple versions of articles, which introduces the third problem—duplication of effort for content creators and consumers.

Chapter 2
Crowdsourcing Information Systems

2.1 Overview

Crowdsourcing information systems can be classified in many different ways. One of the classifications can be the nature of collaboration: explicit or implicit. In explicit collaboration systems (e.g., Wikipedia or Linux), users collaborate explicitly to build information artifacts. On the other hand, implicit collaboration systems let users collaborate implicitly to solve a problem for the system owners. For instance, the ESP game [33] makes users collaboratively label images as a side effect while playing the game.

Second type of classification can be based on the type of the target problem. The target problem can be any problem defined by the system owners, from building temporary or permanent information artifacts to executing tasks. Another dimension can be the degree of manual effort. When building a crowdsourcing system, system owners must decide how much manual effort is required to maintain the system. This can range from relatively little (e.g., combining ratings) to substantial (e.g., combining code), and also depends on how much the system is automated. The system owners must decide how to divide the manual effort between the users and themselves. Some systems ask the users to do relatively little and the owners a great deal. For instance, to detect malicious users, the users may simply click a button to report suspicious behaviors, whereas the owners must carefully examine all relevant evidence to determine if a user is indeed malicious. Some systems do the reverse. For example, most of the manual burden of merging Wikipedia edits falls on the users who are currently editing, not the owner.

The fourth criteria for classification can be the role of human users. Here, we can consider four basic roles for humans in a crowdsourcing system. *Slaves*: humans help solving the problem in a divide-and-conquer fashion, and minimize the resources (e.g., time, effort) required by the owners. Examples are ESP games and finding a missing boat in satellite images using Mechanical Turk-based systems [154]. *Perspective providers*: humans contribute different perspectives that when combined produce a better solution than with a single perspective. Examples are reviewing books and aggregating user bets to make predictions [195]. *Content providers*: humans contribute self-generated content (e.g., videos on YouTube or images on

A. Ranj Bar, M. Maheswaran, *Confidentiality and Integrity in Crowdsourcing Systems*, SpringerBriefs in Applied Sciences and Technology, DOI 10.1007/978-3-319-02717-3_2, © The Author(s) 2014

Flickr). *Component providers*: humans function as components in the target artifact, such as a social network, or simply just a community of users (e.g., the owner can sell ads). Humans often play multiple roles within a single crowdsourcing system (for example, slaves, perspective providers, and content providers in Wikipedia) [80].

2.2 Major Crowdsourcing Systems

We introduce the most widely used crowdsourcing information systems by categorizing them into four different groups based on the target problem. For this categorization, we use collective knowledge management, collective creativity, collaborative gaming and collaborative voting as the four groups. Because online social networks are one of the most important class of crowdsourcing information systems, we discuss them separately in the next section.

2.2.1 Collective Knowledge Management

This type of systems allows users to build artifacts often merging user inputs tightly and requiring users to edit and merge one another's inputs. A well-known artifact created by this type of system is textual knowledge bases (KBs). To build such KBs, users contribute data such as sentences, paragraphs, web pages, and edit and merge one another's contributions. The two main examples of crowdsourcing KB systems are Wikipedia and Yahoo! Answers.

Wikipedia is an online encyclopedia that is freely available. The notion of open editing in Wikipedia encourages many people to collaborate in a distributed manner to create and maintain a repository of information artifacts. Wikipedia has more than 17 million registered authors and more than four million articles [2]. It has become a valuable resource and many people cite it as a credible information source. However, the open process that provides popularity to Wikipedia makes it difficult for readers to ascertain the credibility of the content. Similar to other crowdsourcing systems, Wikipedia articles are constantly changed by contributors who can be non-experts or even vandals. On the other hand, Yahoo! Answers is a general question-answering forum to provide automated collection of human reviewed data at Internet-scale.

2.2.2 Collective Creativity

The role of human in creativity cannot be replaced by any advanced technologies. The creative tasks such as drawing and coding can only be done by humans. Here, crowdsourcing is used to tap into online communities of thousands of users to develop original products and concepts in areas such as photography, advertising, filming, video production, graphic design, and apparel design. As a result, several entrepreneurs have setup crowdsourcing systems (e.g., Sheep Market) to harness the power of the crowds to cheaply complete creative tasks. The Sheep Market is a web-based artwork to implicate thousands of online workers in the creation of a massive database of drawings. It is a collection of 10,000 sheep created by MTurk workers, and each worker was paid US$ 0.02 to draw a sheep facing left [3, 121].

Another example is Threadless which is a platform of collecting graphic T-shirt designs created by the community. Although technology advances rapidly nowadays, computers still have no clue about how to creatively solve a specific problem to develop a new product. In Threadless, different individuals may come up with different design ideas for T-shirts [55] from which the most appropriate ones could be selected. Moreover, Leimeister [130] proposed crowdsourcing software development tasks as ideas for competitions to motivate user participation in crowdsourcing. Well known software such as Apache, Linux, Hadoop are produced and maintained by crowdsourcing systems.

2.2.3 Collaborative Gaming

The concept of *games with a purpose* was pioneered by Luis Von Ahn and his colleagues [32]. The games produce useful metadata as a by-product. By taking advantage of people's desire to be entertained, problems can be solved efficiently through online games. The online ESP Game [33] was the first human computation system, and it was subsequently adopted as the Google Image Labeler. Its objective is to collect labels for images on the Web. In addition to image annotation, the Peekaboom system [34] can help determine the location of objects in images and provide complete outlines of the objects in an image. The concept of the ESP Game has been applied to other problems. For instance, the TagATune system [128], MajorMiner [138] and TheListen Game [206] provide annotation for sounds and music which can improve audio searches.

2.2.4 Collaborative Voting

In this type of crowdsourcing systems, a user is required to select an answer from a number of choices. The answer that the majority of the users select is considered to be correct. Voting can be used as a tool to evaluate the correctness of an answer from the crowd. An example of popular crowdsourcing websites with collaborative voting is Amazon Mechanical Turk (or MTurk) [4]. A large number of applications or experiments are conducted in Amazon's MTurk site. It can support a large number of voting tasks.

2.3 Social Networks

Social networks are social structures that consist of social entities (e.g., individuals, groups, organizations) that are connected to one another by social relationships [215]. Social relationships can be quite broad; examples include friendships, behavioral interactions, biological relationships, or affiliations.

Social network analysis focuses on studying the different patterns of relationships among social entities along with their implications on the behavior and decisions of the social entities [211]. Based on the same concepts used in graph theory, a social network is represented by a graph consisting of a set of nodes and edges. The nodes in the graph represent the social entities, while the edges represent the social ties that link those entities. The resulting graph structures are often complex where the social entities are considered interdependent rather than independent units. This means that in social network analysis the discrete unit of analysis is the combination of social entities and the relationships among them.

Social network analysis has been widely used in recent decades in such diverse areas as sociology, anthropology, biology, economics, and information science [152, 46, 103, 205, 213, 149]. For example, in the area of epidemiology, social network analysis has been used to study the effect of different patterns of social contacts on the spread of human diseases and viruses [152], and also to study the relationship between social and community ties and mortality among people [46]. In the field of sociology, social network analysis played an important role in understanding how information spreads on social networks [103] and how individuals are connected in the physical world [205, 213]. In economics, the influence of social structures on the outcomes of the labor market are analyzed using the tools of social network analysis [149].

The last few years have witnessed the emergence of the second generation of the world wide web—"Web x2.0." By facilitating online collaboration and information sharing for people, Web 2.0 has enabled the development and evolution of online based social networks (communities). This has resulted in an increased use of social network analysis to study the underlying structures of these communities and address the problems and challenges that arise within these online systems. Due to their importance, we introduce online social networks and discuss the different properties associated with them in the next section.

2.4 Online Social Networks

Online social networks are online communities of individuals who share common interests or activities. There are many web portals on the Internet that offer the facilities to create manage online communities with different modalities to socialize among the members. Boyd [54] defines today's social networking websites as "web-based services that allow individuals to (1) construct a public or semi-public profile within a bounded system, (2) articulate a list of other users with whom they share a connection, and (3) view and traverse their list of connections and those made by others within the system."

While the definition for social networking websites provided in [54] presents those sites as being mainly profile-based, there exists many social networking sites that offer other types of services. The research report in [5] presents a much broader categorization of the social networking services that exist today. It identifies eight main types, among which are the popular profile-based social networks like myspace.com and facebook.com. Content-based social networks are also among the most popular sites, where the main form of interactions and relationships between users are established through the creation of user content. Examples of such sites include flickr.com, a photo-sharing site, youtube.com, a site for sharing user created videos, and delicious.com, a social bookmarking and tagging site. In addition, other social networks provide micro-blogging services, where the users post status messages allowing other people on their social network to track their status; an example of such a service is twitter.com. What makes these social networking sites interesting is that they eliminate the physical limitations of the traditional social networks, allowing their participants to extend and build their personal social networks by meeting new individuals from across the globe. As a result, we are witnessing the rise of new and different relationship structures that are not related to the offline world.

Some suggest that online social networking can be traced back to 1997 with the launch of the first blogging site [6] and social networking website sixdegrees.com [54]. Since then, the number of social networking sites has increased dramatically, attracting many users and generating high web traffic. Based on the information provided by Alexa (alexa.com—a database of information about sites that includes various statistics), many of the existing social networking websites are ranked in the top 500 websites in terms of traffic generated on the web [7]. Reports from Nielsen Online [8], a company that provides measurement and analysis of online audiences, indicates that nearly half of the biggest social networking sites are also among the fastest growing, with still room for potential future growth.

Chapter 3
Online Identities

3.1 Overview

Identity can be considered as a certificate or a license that one must have in order to be included in an organization or to be entitled to certain rights. It usually consists of a set of personal information to distinguish the owner from others. The most common identity in real life is the ID cards for citizens of a country. The purpose of this ID card is to claim that one belongs to a country and is a legal resident; hence, she has certain rights by law. However, the concept of identity dates back to ancient times, long before constitutional law was developed. The starting point of identity was when human first differentiated among themselves and started recognizing "self." But identity has become more than just a tool to distinguish people. Akerlof and Kranton include identity into economic models of behavior and find that identity can affect individual interactions [38]. Eaton, Eswaran and Oxoby argue that we need identity or rather an identity system for the sake of survival [84].

To create an identity document, a name must be registered followed by one's birth date and a lot of other information. In some countries, even finger prints are collected to issue a passport. One might think that his name, birth date, home address and which company he's working for and other information are of no use to any other people and won't bring any benefit to a thief. However, there are many opportunities for a thief to profit from those information. For instance, lot of people receive email spam and phone spam. According to the Message Anti-Abuse Working Group, in 2010, email spam took up 90 % of total emails. The annoying spam is just one of the reasons why individuals need to protect their personal information. Sometimes, a person reveals her contact information to a party that she trusts, such as a bank or a bookstore. A staff working at the bank or bookstore can collect a lot of customers' information and can sell it to another company that wants to promote their products. This is only the simplest way individuals' information can be of benefit to third parties.

Identity theft and fraud Identity theft or impersonation refers to the act that some-one claims to be someone else. For instance, a thief can create a fake profile by copying information about a person and then use the social connections to defraud

A. Ranj Bar, M. Maheswaran, *Confidentiality and Integrity in Crowdsourcing Systems,*
SpringerBriefs in Applied Sciences and Technology, DOI 10.1007/978-3-319-02717-3_3,
© The Author(s) 2014

trusting friends of the victim. Another example can be where someone claims to be a representative of a corporation or a well-known company. Here, the thief can claim that he has inside information about the target corporation which can be leveraged by the unsuspecting victims in their dealings related to the corporation. This can be very similar to spoofing attacks on computer networks. As these kinds of identity thefts have been reported a lot, people are becoming more aware of them. The most popular illustration of identity theft might be the movie Catch Me If You Can, which is a movie based on the true story of Frank Abagnale Jr. who assumed more than eight identities to forge checks and escape from U.S. custody. After serving five years in prison, he became an American security consultant. Despite its various forms, identity theft has one common goal, that is, to escape from punishments, especially on the Internet. Thieves can make use of it because people decide to trust a specific person based on her identity.

Identity is not a trivial issue on the Internet. With the boom of Internet technologies and all sorts of Internet-based companies, identity is not just a way of only differentiating people. It affects the extent people trust each other. One of the main advantages of Internet is freedom, freedom of speech, freedom on meeting strangers from all over the world or freedom of choosing your character. Currently on the Internet, individuals can create any user name (called a pseudonym) and hide behind it. Unlike the ID in real life, pseudonym is easy to create and disown. In a lot of online games, people often create more than one account and use these accounts to help each to level up faster. However, this freedom is not only available to good people, but also to thieves. To catch potential victims, thieves try to make fake identities on the spot. They can renew their online identities many times in a short period of time. The result is a new type of Internet scam called whitewashing [75]. This scenario mostly occurs on online stores and e-commerce websites. According to Javelin Strategy and Research and cited by the U.S. government [23], in 2011, 8.1 million U.S. adults were victims of identity theft or fraud, with a total cost of approximately 37 billion dollars. In 2008, the average out-of-pocket loss due to identity theft was 631 dollars per incident.

Protection against identity threats One might ask what if online shopping websites have a strict identity creation so that would make it harder for an individual to create multiple accounts. While such a design can slow down whitewashing attacks, there are other reasons for not using this approach. With fixed identities, it becomes easier to track one's online activities. Imposing single ID per user slows down whitewashing attacks; however, users face new threats from information tracking or leakage. One of the advantages of pseudonym is that it is easy to create. Fixed identity is usually based on real-life credentials and users might not want to give that information just to buy a pair of shoes online. Also if users are releasing these credentials online, with phishing attacks becoming more sophisticated, there's an even higher risk of identity theft and fraud. In 2004, the U.S. government developed the Federal Employee ID Card. This ID card is a smart card and contains information such as the employee's name, agency, finger prints, identification numbers and certificate to access different systems. Later on, the REAL ID Act of 2005 [15] proposed national ID cards for the public while privacy advocates have serious reservation regarding Real ID such as the database storing personal information of the whole country getting stolen. Another concern is that it will be too easy for the government to track its citizens. These concerns stopped the creation of the US national ID. Because identity is an important issue, we discuss online identities in more details. Online identities can be categorized into three types: fixed identity, pseudonym and social identity.

3.2 Fixed Identity

Fixed identity is exactly what we have in real life. Everyone has one and only one identity and no two persons have the same identity. Among the three types of identity, fixed identity is the most distinguishing one, mostly because its original purpose was just to distinguish people. Identity was originally invented in the fixed form. Social security number, driver's license and passport (within a country) can all be considered as fixed identities as well as URL addresses which can be seen as identities for resources.

3.2.1 Online Fixed Identity

It is not easy to find an example of a website using fixed identities. However, there are several websites that limit the number of accounts a user can register. One of the terms in PayPal's user agreement prohibits a user from holding multiple accounts [17]. PayPal has the right to terminate multiple accounts that belong to the same user. Another website, NeoBux [18] which pays users to view the advertisements they show also has a policy that allows only one user per IP in a 24 h period and only one account per computer [19]. A user can register for multiple accounts but she needs to have the same number of computers. This is more like a fixed identity based on IP addresses. With the time limit of 24 h, it is possible for another user to use the same computer at a later time. In addition, when Facebook first started, its accounts were based on credentials (emails) issued by educational institutions. Each user had to be enrolled in a school to have an active email. Because most schools institute a one email per student policy, users were restricted to only one account on Facebook as well—which has been relaxed in recent years.

3.2.2 Advantages and Disadvantages of Fixed Identity

The advantage of fixed identity is its security, as long as the authentication process can guarantee that a user is really who he claims to be. The disadvantage is its centralization and possible personal information leakage. When a fixed identity is created a lot of unique identity attributes are attached to the identity to ensure that the online identity is well connected to the offline identity of the user. Therefore, an online fixed identity depends on other real life fixed identities. With the constant threat from hackers trying to steal user information, authentication processes will become more and more complex for fixed identities. Chaum argues that fixed identities provide one-sided security, protecting service provider from individual users while users' information are left in danger [67]. One approach to this problem is to limit the scope of the fixed identities such that there is one per service. One implementation of such a scheme is using a card computer as an intermediary between users and service providers. Thus, only users know about their information and service providers authenticate the card computer with the pseudonym. In this way, the intermediate card computer provides multiple pseudonyms for the user and service providers cannot link the pseudonyms.

3.3 Pseudonym

Pseudonyms are the most commonly used type of identity on the Internet. We can categorize its creation into two levels based on how much information a user has to provide. The easiest way to create a pseudonym is to provide a name, a password and maybe a security question. The name is not necessarily a real name. It can be any name for an account which is visible to other people on the site. Another way to create a pseudonym is to provide name, email address, a password and maybe birth date and gender. Facebook lets users to have an account in their site by providing such information. Services that consume email addresses to provide pseudonyms could access the email box of the user and reach more people via the user's contact list. Also, by using the email addresses, the services can ascertain whether there is a human user at the given email address.

3.3.1 Organizing Online Pseudonyms

Most of the pseudonyms used on the Internet are based on email addresses. However, some are based on other pseudonyms such as Mashable (http://mashable.com) and Zoho (http://www.zoho.com), which are based on users' Google accounts. Flickr (http://www.flickr.com) lets users to login with their Facebook accounts. Websites that accept OpenIDs (http://openid.net) allow users to sign into their sites using credentials issued by the OpenID issuing site [167]. The OpenID credentials look like an ordinary URL and is human friendly. It saves users a lot of time that would otherwise be spent on creating and managing account names and passwords on diverse sites. Mainly, OpenID links pseudonyms or reduces pseudonyms. The disadvantage is that if a hacker gets a user's OpenID, he can use the user's identity on all sites that rely on OpenID credentials. In addition, the OpenID can be "phished" in more websites. Oh and Jin discuss the limitations of OpenID and show that hijacking attack during the authenticating session is possible with OpenIDs [21].

3.3.2 Advantages and Disadvantages of Pseudonym

The benefit of using pseudonym is its simplicity. Thus, pseudonym makes it easier for users accessing websites and their services. This also helps the thriving Internet-based startups. Another advantage of a pseudonym is that it protects the real identity of its owner. It is similar to wearing a mask on the Internet. People can enjoy taking multiple personas to suit their needs. For instance, a user can create one account for his business life and another for social life. Neubauer and Riedl propose an improved electronic health care architecture which makes use of pseudonym and better protects patients privacy [153].

Accordingly, there is an interesting question: how easy is it for someone to actually hide on the Internet using pseudonyms? There have been several incidents where online pseudonyms have been tracked down to the actual user. This phenomenon is called the "human flesh search engine" during which a big number of netizens search the web for any information regarding actual person behind a pseudonym who might have done something outrageous. The search initiated by the human flesh search engine continues until the true identity is found. The beginning of this "human flesh search engine" was in 2006 when netizens were searching for the real identity behind a pseudonym, Poison. Poison posted a lot of pictures showing off luxurious life style on an online forum called Tianya (http://www.tianya.cn). Generally, the "human flesh search" is a way for the people to use the Internet to solve a technical problem. Therefore, pseudonym is not a perfect mask and if a lot of people want to find out about a specific user, they can tear down the mask. In [192], Stutzman states that information released on a social network such as photos posted, political views expressed, and education could also be used to identify people and should be given more attention by the users.

Another way to uncover the real identity behind a pseudonym is to get acquainted with the person behind the pseudonym and determine the true identity. In 2005, a college student tried to find someone to write her school paper on the Internet. A comedy writer responded to her post and offered to write her paper. After talking to the student, the comedy writer released the real identity of the student, which brought huge dispute whether what the comedy writer did was appropriate. Therefore, using pseudonym only gives users a limited privacy. Privacy is achieved by hiding the real name while releasing associated history actions and other information such as purchasing history or browsing history. However, if users hide their traces on the Internet and release their names, it may not be a risky action. Sweeney proposes a k-anonymity model that mixes k users together when a user is submitting her information, so that no one can tell which user in this k users group released a particular piece of information [198].

There is yet another privacy issue concerning pseudonymous systems. In [158], Pashalidis and Meyer show that it is feasible for collaborating organizations to link the transactions of users in a pseudonymous credential system. Same type of concern is also mentioned in [68]; Chen and Rahman surveyed the privacy designs of mobile SNAs (social networking applications) and concluded that in most of them there is a little information provided on the level of protection afforded to the information after they are submitted to the backend servers. Pashalidis and Mitchell identify limits to pseudonym unlinkability in [158]. In their work, they show how the timing attack can break pseudonym unlinkability by identity issuing and verifying organizations working together to figure out which pseudonym belongs to which user. Clearly, what pseudonym provides is security against other users but not against the systems that are involved in issuing and verifying the pseudonyms.

One of the disadvantages of using pseudonyms is that they can be easily abused. It provides malicious users with the same amount of safety as the legitimate users. Unlike legitimate users, malicious users are always changing their accounts, making it almost impossible for websites or other users to track their activities. However legitimate users who stay with one or just a few accounts can be tracked. While the legitimate users keep their pseudonyms unchanged to gather sufficient reputation, malicious users change their pseudonyms so that they could maximize the profit they gain by cheating others. Recently, more and more users are aware of releasing personal information on the Internet and they are more reluctant to interact with strangers. Krasnova et al. define two types of threats that current OSN users are facing: organizational threats and social threats [124].

Identity theft also can happen with pseudonyms. Marshall et al. studied online identity thefts [141]. In their work, they point out methods of online identity theft: protocol weakness, naive users, malicious software, data acquisition and network impersonation. They discover that pseudonym makes it just easier to steal someone's identity on the Internet. Bonneau et al. showed that with just limited information revealed about each user, many properties of the whole social graph can be even reverse engineered [53]. He et al. showed that personal attributes can be estimated especially for people who have strong ties (closely related) with other people [107]. Even if people choose not to disclose their private information, it can still be inferred. Furthermore, these identity theft attacks can be automated. Bilge et al. showed how easy it would be to launch automated crawling and identity theft on popular social networks [47].

To solve these problems, it was suggested to use long term reputation as measure for trust between users. However, Friedman and Resnick present another problem related to newcomers. New and innocent users are not trusted any more as the social cost of using cheap pseudonyms [92]. Same problem exists on P2P systems. Adar and Huberman present free riding problem on Gnutella system [26].

Another type of attack on pseudonyms is Sybil attack, where a malicious user creates a large number of accounts on a website and uses them to outnumber legitimate users. Levine et al. survey several examples of Sybil attacks and their corresponding solutions [133]. For instance, companies can use Sybil attacks to increase Google Page Rank rating to boost up their rank in search results. Douceur studied Sybil attacks in large-scale P2P systems and points out that most solutions rely on certain assumptions that are either unjustifiable or unrealizable [83]. Yu et al. leveraged social networks to defend against Sybil attacks [221]. Their basic idea is that when there are a lot of Sybil nodes on social graphs, cutting a small number of edges will isolate a large number of Sybil nodes whereas this would not happen with normal nodes. Other than Sybil attack, with pseudonyms, people can share accounts in an unauthorized manner. For example, friends can share iTunes account to avoid paying for music. Fortunately, attention have been paid to this problem. Lysyanskaya developed a theoretical construction for a pseudonym system that discourages people from sharing accounts [136].

Some researchers argue that because of pseudonyms, people think it is fine to be dishonest on the Internet. If this is true, we have a cyclic situation where pseudonyms give us the protection and in return, we like it more than fixed identities because we can be irresponsible with our online activities. To this end, giant social networking sites such as Facebook have a policy that urges users to provide truthful information when registering for accounts. In Facebook, "truthful information" includes no fake name, accurate contact information and no false personal information. Under the Facebook Registration and Account Security section, there are rules which require users not to create account for other people, not to create more than one account and also not to transfer their accounts. Moreover, Facebook reserves the right to shut down an account if any of its rules are breached by the account holder. One of the famous incidents of account takedowns relates to "John Swift." John Swift was originally the name of a famous philosopher and writer. The name was used by a blogger on Facebook. Facebook deleted the account and replied to the owner questioning about the deletion of his account as: fake account is an instance of the violation of Facebook terms of service. Another example was the account registered under a pseudonym of Michael Anti's. The owner was a Chinese journalist who used the name on published articles, blogs and even Harvard fellowship documents. Because the name was not the real name of the user, Facebook blocked the account. With its real name policy, Facebook is trying to maintain a more accountable Internet environment.

Other social networking websites are working toward a similar direction as the one instituted by Facebook. PatientsLikeMe (http://www.patientslikeme.com), for instance, collects user information and sells it to medical and drug companies, helping them develop new products. Patients can share their medical experiences and compare them with other patients suffering the same diseases. Obviously, users of PatientsLikeMe prefer real patients and not fake ones. Therefore, this site has a rule which forces users to provide truthful and accurate information. Users in eBay site are also required not to misrepresent others.

No matter how much Internet companies try it is difficult to make people act honestly all the time in the Internet. Ordinary users are often not willing to register with their real name fearing privacy. Pseudonym is now in the middle of an awkward situation where some people want to replace it with fixed identities to avoid all the problem with it and others prefer pseudonyms to enjoy all the benefits provided by it.

Various efforts have been made to address pseudonym's low accountability. In [159], Penna et al. showed that having reputation associated with pseudonyms is better for P2P systems than relying on incentive schemes. Ford and Strauss proposed pseudonym parties to ensure everyone has only one pseudonym while maintaining user's anonymity [91]. Users need to go to "pseudonym parties" on "Pseudonym Day" and get a certificate in person together with a hand stamp to prevent the user from getting a second certificate. With the certificate, the user can then register with pseudonym servers. Whenever the user needs to provide her identity to a relying website, the site would redirect the user to the pseudonym server which would authenticate her. After the user is verified by the pseudonym server, she will return to the relying website and register for an account on the website. There is no need for the user to use the same pseudonym in every relying website. Only the pseudonym server knows about each of the pseudonyms a user owns on each relying website. This is very similar to OpenID, which links different pseudonyms of one user and protects the connections.

Because of the widespread use of pseudonyms on the Internet many problems related to them have been discovered. For instance, Yokoo et al. studied the false-name bids on auction sites [219]. False-name bids are defined as bids submitted by a single user but through different accounts. In their study, they showed that there is no false-name proof auction protocol which is Pareto efficient and everyone gets their optimal benefit. Another example is that on websites like Twitter (http://www.twitter.com), Google+ (https://plus.google.com) and Weibo (http://weibo.com) users can "follow" friends, celebrities or even strangers. These websites let users to post short messages, videos or pictures to their followers. For marketing purposes, accounts with large numbers of followers are very valuable. This has caught the attention of profit seekers who register few accounts and post useful tips to other users, including news, photos, or anything to attract more "followers." When they have aggregated enough "followers," they can sell their accounts to companies who want to promote themselves. There are also scenarios where all the "followers" are fake accounts. Since pseudonyms are so easy to get, anyone can create a lot of accounts in a short time. A website itself can actually produce a great number of dummy accounts and use them to boost the value of the website for advertisers.

Pseudonym was first applied on the Internet due to its simplicity and ease of deployment. With all the problem surrounding the use of pseudonyms it might be a suitable time to reinvent identity on the Internet.

3.3.3 Whitewashing

Pseudonyms make whitewashing possible. Whitewashing refers to malicious users cheating on others and rejoining the online community under new identities to start their attacks all over again. It happens in several different types of websites. In P2P systems, peers are supposed to upload media objects for other users while getting data they need from other users. Because media sharing is voluntary, users can decide not to upload any file and just get files from other users. These users are recognized by the system as free-riders. Usually systems impose penalties on free-riding. Because P2P systems are using pseudonyms, free-riders can get away with a new identity and come back afterwards. This way, a user becomes a whitewasher and can keeps free-riding.

On online shopping websites, whitewashers first cheat on some victims and get new identities to rejoin the website and cheat on others. Because whitewashing can adversely affect the overall shopping experience of other users, strict measures are in place to slow down this problem. One common way of handling whitewashing is to use fixed identity such as a credit card number to anchor a pseudonym.

Whitewashing can be rampant in recommendation systems. It can be used to boost sales of a company's products or to diminish competitors' profits. There is a business motive behind this type of whitewashing. Because people often go to sites like Netflix (http://www.netflix.com) or IMDB (http://www.imdb.com) to see comments about a movie they want to watch or sites like Goodreads (http://www.goodreads.com) or aNobii (http://www.anobii.com) to check reviews of a book they want to buy, companies could hire people to post flattering about their products in these sites. In addition, someone can just do that for fun because pseudonyms on rating sites are free anyway. Obviously, both false positive and false negative comments could harm the credibility of a rating website.

3.4 Social Identity

As online social networks become popular, the concept of social identity is gaining traction. Social identity is explained in many different ways depending on the discipline under which it is investigated. In psychology, social identity refers to an individual's self-perception due to being a member of particular social groups [16]. Websites like Facebook has taken this definition into computer networks and provides a platform that connects people just like how we get to know each other in real life. A lot of applications have been developed to leverage these connections created on social networks. They can also be utilized for identifying people. Here, the social identity we are talking about is not just a Facebook profile but a digital identifier generated based on the social graphs that social networks such as Facebook have collected.

3.4.1 Making Use of Connections

There are different forms of virtual connections in the Internet and they can be leveraged in different ways too. Facebook connections, email contacts, and MSN contacts are examples of social connections whereas RSS subscriptions and even bookmarks are examples of people-content connections. For example, a social networking service could exploit the email contacts of a newcomer by automatically inviting the newcomer's friends to its portal. A social networking site will become valuable to a person if most of her friends are already there.

Social connections can also be used for user authentication purposes. Brainard et al. proposed using friendship links to authenticate a user [56]. When a user is not able to authenticate herself using the primary authentication mechanisms (e.g., forgotten to bring the authentication token in a two-factor authentication), she can ask a friend to vouch for her. The "helper" can create a temporary vouching token and issue it to the "asker," who can use it temporarily for authentication purposes with another factor such as a username and password. Soleymani and Maheswaran has proposed a scheme that puts Brainard's fourth factor in use when authenticating users are in the mobile social context [186]. Their scheme assumes that a user is always carrying around her mobile phone and the mobile phone is a witness for all the social interactions of the user. In addition, the mobile phone is able to vouch for interactions in the cyberspace as well (e.g., email or instant messaging). Using the secure tokens issued for witnessing the interactions, the mobile phone can vouch for the identity of the user.

In addition to individual social links, the topological structure of social graphs can be used to defend against Sybil attacks. In [208], Viswanath et al. investigated existing social network-based Sybil defenses. Despite considerable differences, all existing Sybil defense schemes depend on community detection and are based on assumptions that Sybils can only form a few connections to non-Sybils and that the presence of Sybils lead to misbehavior. Viswanath et al. discovered that the reliance on community detection makes existing Sybil defenses vulnerable.

Regarding the identity theft we mentioned previously, Jin et al. have characterized the behaviors of identity clone attacks and devised ways to detect suspicious identities in [111]. They compute similarities between profiles by taking into account the friend lists and attributes of a profile. However, in the case of whitewashing, where users are not necessarily building profiles and adding friends, detecting whitewashers might not be effective. Because normal users may have poorly maintained their profiles, it is difficult to distinguish between legitimate users and whitewashers just from the attributes and friends list. Once a profile is detected as a fake one in an OSN, any account related to this profile on other networks can be labeled as whitewashers and this can prevent whitewashing.

The topological structure of a social network can impact the searchability, information dissemination, and collaboration among users. In [212], Watts et al. offered an explanation for social network's searchability. Mislove et al. gave the structural measurements of online social networks in [145]. They found that online social networks have special advantages in disseminating information and inferring trust among users. Subramani and Rajagopalan designed a framework for viral-marketing, where interested consumers can tell each other about a product and the information spread is based on the structure of the online social network [194]. Leskovec et al. even used online social networks to predict people's attitudes towards others based on their connections with their friends [132]. Kimball and Rheingold presented how online social networks can help big organizations to get organized and to communicate efficiently and hence reduce their internal friction [119]. Mankoff et al. tried to encourage users to reduce their ecological footprints which are expected to have a large impacts on users [139]. Cachia et al. claimed that online social networks can foster creativity, reflect changes in social behavior and accumulate collaborative intelligence [61].

3.4.2   Generating Social Identity

We can generate social identities by naming the social links and using the unique ways people are linked to each other to name people in network. Maheswaran et al propose a framework to generate social identities for users [137]. In this framework, relying parties can use social identities for authenticating users. Each user in a social network can be considered as a node and their connections are links between the nodes. Therefore, a social network can be viewed as a connected graph. Social identities are generated from the view of several landmark servers that are strategically linked to chosen nodes in the social graph. From the perspective of the landmark server, the nodes are divided into different sets based on their hop distances to the closest landmark server. A user's social identity consists of a set of edge-disjoint paths from the landmark servers.

A user-centric identity selection system such as Windows Cardspace can be used to manage the social identity of a user. The user can register with a social identity provider to obtain her social identities. The social identity provider is responsible for profiling the social network neighborhood of the user and generating the appropriate social identity tokens. Every time the user needs to be authenticated, the identity selection framework running on user's device will determine the minimum requirements on the identity and the interact with the social identity provider to obtain an appropriate social identity token as shown in Fig. 3.1. The social identity provider calculates the latest social identity factoring in newly created or deleted connections. Heartbeat messages are sent regularly throughout the social graph to detect new or lost connections. The user's identity selection framework can issue this social identity token as proof of identity to a relying party. The relying party can verify the authenticity of the token and deduce the social connectivity of the issuing user. The social identity framework can restrict the release of connectivity information based on the privacy settings of the issuing user.

3.4.3   Properties of Social Identity

In this section, we discuss the properties of social identity that make it different from fixed identity and pseudonym. Social identity is based on connections one has with other people but not based on local credentials of the user. The users need not

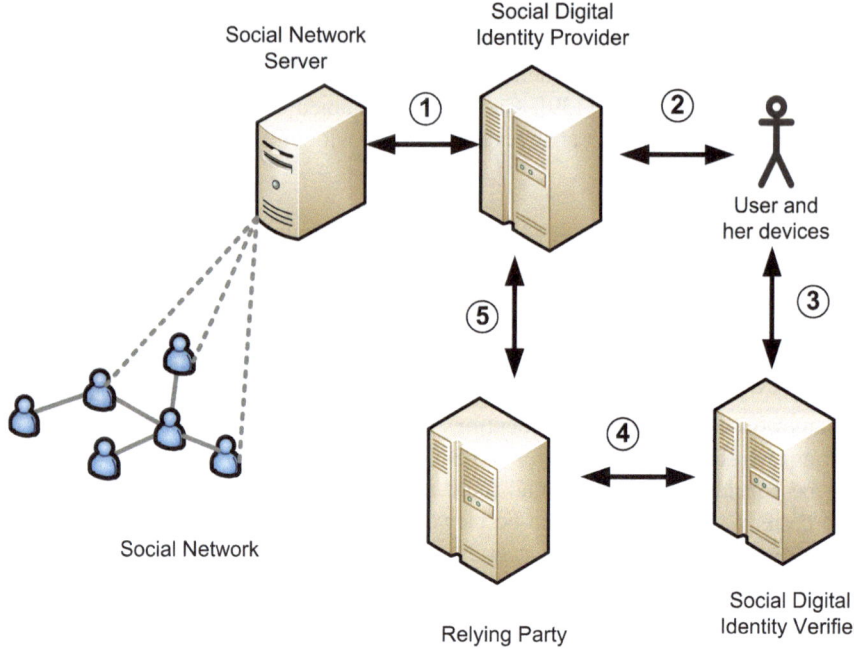

Fig. 3.1 An example deployment scenario for social digital identities. [137]

reveal any privacy sensitive information to establish their identities. For example, users need not attach sensitive attributes such as social security number, driver's license number, or household registration. Even, there is no need to register with real name. Social graphs are maintained by social network providers. While the local changes in social graphs are known to the users, global changes are only known to the social network providers. With social identities, it is hard for malicious users to forge identities.

Regarding user privacy, when the user provides a social identity to a third party, the third party cannot have access to information related to the social identity. The third party can only send the social identity to a social identity provider and get answers about the validity. Therefore, social identity protects a user's real identity better than fixed identity. Before issuing the social identity tokens for a user, the social identity provider challenges the user to establish the ownership to a node in the social graph. This way a user is prevented from impersonating another user in the social graph. However, a user can have multiple accounts, for instance, one for work, another one for personal relations. The user need to select which identity among the ones she posses that should be used in a particular scenario. Once the identity is issued, the relying party will validate it by contacting the social identity verifying server.

Unlike pseudonyms, it is not necessary to generate separate identities for the different sites a user might visit. A single social identity can be shared among the different sites. However, for privacy reasons the linking among the different identities can be restricted through the social identity provider. The websites do not need to build their own identity management systems, instead they could just be relying parties on the social identity providers. Thus, the Internet can become a virtual community and users can decide whether to trust someone similar to situation in real life. Because social identities are persistent, a user can have reputation even when she is a newcomer in a particular site—reputation ported from prior activities in other sites.

The essential part about social identity is the interactions between social networks, social identity provider and identity selection frameworks. Social networks provide the data which is necessary to generate a social identity for a user. There can be multiple social networks feeding data to a social identity provider. They need to guarantee that the social graph is secured and will not be released to any third party. Here, a user is the one who asks for issuing the social identity. Also, the user is the person who is responsible for releasing her social identity to other parties. Social identity puts users in-charge of their credentials and personal information.

3.4.4 Applications of Social Identity

As pointed out in [72], the business world has adopted the concept of social networking and there is a growing need for a unified digital identity source. Social identity is not merely a name but encapsulates the reputation of the owner of the identity as well. It hides the real identity of the owner from organizations and provides organizations with a convenient way to authenticate the owner. Therefore, social identity protects both parties. Because social identity is based on people's relations, we can design better trust models if we categorize people's relations into different trust levels.

Jennings and Finkelstein proposed a trust model that can help in constructing community-aware identity management systems [110]. Agarwal et al. presented the long tail distribution of social networks: most people have few contacts and few people have a lot contacts. They showed how social graphs can be used to find "familiar strangers," people who do not know each other (not friends) but are familiar with each other [29]. In addition, they showed that by more accountable virtual community, we can enlarge the virtual community in order to have more reliable Internet. For instance, when a user tries to do groceries online, she can first send the shopping list to the store online. Afterwards, the store should verify that the user is a valid resident in its service area through a social identity provider. By passing the validation phase, store can prepare the grocery and the user can pick it up at the end of the day.

3.5   Discussion

Identity is a challenging issue. It decides how we interact with each other and how much we can trust other people on the Internet. With an ideal identity system, a user can decide to reveal or hide her information and to whom to reveal her identity as well as to what extent she disclose the identity. A proper identity management scheme should enhance security of organizations and companies without compromising users' privacy. Recently, online identities and social networking sites showed an interesting connection between them. Despite many technical proposals over the last decade, implementing socially acceptable online identities remains an open problem. The social networking, on the other hand, has flourished on the Internet and is leading the transformation of the web into a social web. The online implementations of social networks provide the tools for users to create online profiles and link with other people. In many ways these online profiles, data associated with the online profiles, and the connections formed with the profiles can be considered as online identities of people. However, there is no acceptable framework or methodology for validating the information that is available on online social networks. As a result, the level of assurance that can be placed by the online identities developed on online social networks remain low.

Chapter 4
Confidentiality Control

4.1 Overview

The privacy and security concerns in a crowdsourcing information system are dependent on how the ownerships of the content created by the system are managed. A crowdsourcing information system can use different ownership models: (a) creator owned, (b) system owned, and (c) group owned. In creator- owned systems, the creator of a data object is responsible for setting the access control constraints for the data object she creates. Examples of crowdsourcing information systems that use this model are online social networks and blogs. In system-owned crowdsourcing systems, individual contributors do not have the privilege of setting the access control policies for the content they create as part of their activities. No matter who generates the content it is owned by the system and managed based on its policies. Wikipedia is an example of system-owned crowdsourcing information system. There are crowdsourcing systems that allow groups of people to collaboratively engage in the creation of information artifacts. In such systems, the data objects are collectively owned by the groups. The members of the groups or the administrators of the groups are responsible for setting the access control policies. Wikis setup for organizations follow a group- based access control model, where members of specific groups own certain portions of the wikis. With decentralized but still controlled ownership the content held by the wikis could be more effectively managed to reflect the goals of the organization.

In addition to ownership, information flow in a crowdsourcing information system could be affected by other factors. For instance, some crowdsourcing information systems can organize their solicitations for contributions as competitions (e.g., T-shirt designing competitions). In such competitions, participants can generate the content (i.e., designs in the T-shirt designing activity) in groups or individually. In competitive crowdsourcing situations, during the formative stages of the content generation, the participants might want to secure the content such that it does not leak to opposing teams.

Because crowdsourcing information systems are driven by "human dynamics," certain amount of information exchange is essential to sustain the crowd activities. This tension between secrecy and publicity is quite evident in online social networks.

A. Ranj Bar, M. Maheswaran, *Confidentiality and Integrity in Crowdsourcing Systems,* SpringerBriefs in Applied Sciences and Technology, DOI 10.1007/978-3-319-02717-3_4, © The Author(s) 2014

In this chapter, we discuss how confidentiality of data objects can be controlled in crowdsourcing information systems.

4.2 Privacy and Security Challenges

The primary challenge for information sharing in OSNs is the impreciseness of the problem itself. In a typical corporate computing system, information sharing is dictated by the overall organizational policies, which are formulated based on the corporate agenda. The information sharing problem in OSNs, however, is not governed by a precise policy. The need to socialize in OSNs dictates that users should share information. However, the privacy concerns can reduce the overall information spread in OSNs. Another challenge is the diverse user populations in OSNs. It is generally accepted that users in OSNs desire to have effortless ways of controlling information sharing. While retaining simplicity, users want mechanisms that minimize unintentional release of data. Yet another challenge is the mismatch of goals among the different stakeholders of OSNs. The OSN operators want unhindered information flow so that they could extract sufficient business intelligence on the user population. The users, on the other hand, want to control information flow to suit their needs and privacy preferences. This means that the control of the information sharing should balance the need for privacy with the need for publicity.

Due to its importance, there has been a number of proposed solutions to the problem of online information sharing on OSNs based on different access control mechanisms. Since this book focuses on the problem of preserving the confidentiality and integrity of shared data, we believe it is worthwhile to review the major access control techniques along with some of the solutions treating the problem of online sharing.

4.3 Access Control Schemes

Confidentiality is defined in [49] as "ensuring that information is accessible only to those authorized to have access." Access control mechanisms are an essential part of information security because they provide the necessary means for preserving information confidentiality and integrity. Today, due to the popularity of online information sharing, there is a pressing need for new access control techniques that provide a secure environment for online information sharing. In this section, we first give a brief introduction to the major access control techniques used in security systems then, we review some of the recent work addressing the problem of information sharing in OSNs.

4.3.1 Major Access Control Techniques

There are three major classes of access control policies that have emerged since the 1970s: Discretionary Access Control (DAC) [127], Mandatory Access Control

(MAC) [43], [44], and more recently, Role Based Access Control (RBAC) [87]. These access control policies are the among the most commonly used in computer systems.

1. Discretionary Access Control (DAC) is an owner- centric based policy where the owner of the protected data dictates the different access policies for the data. Many of the implemented access control policies are related to DAC in some form or another [45]. DAC consists of three main entities: the protected objects, subjects, and access rights. In the system, each object is assigned an owner who is initially the creator of the object. The owner of an object has complete control over the access rights and permissions assigned to other subjects in the system. Subjects are granted access only if the access rights authorize them to perform the requested operation.

   Advantages of DAC is its simplicity, flexibility, and ease of implementation [45]. The main drawback of DAC policies is that the access restrictions can be easily bypassed [181]. A subject who has been granted access can easily pass the object to non-authorized subjects without the owner's knowledge. This is because there is no restriction imposed upon the dissemination of information once a subject has gained access. DAC is usually implemented by Access control list (ACL) and capability based access control systems [181].

2. Mandatory Access Control (MAC) is based on the security classifications of subjects and objects in the system. In MAC, subjects are assigned security classification (or clearance) that corresponds to the trustworthiness of that entity, while the security sensitivity label of an object corresponds to the trust level required for the subjects to have access. In contrast to DAC, object owners do not make policy access decisions or assign security attributes [45]. Access is granted only if a subject has the necessary clearance to access an object. MAC aims to enforce lattice-based information flow constraints to establish high assurance information systems [179].

   This is usually achieved through the following two principles [181]:

   - **Read down**: A subject's clearance must dominate the security level of the object being read
   - **Write up**: A subject's clearance must be dominated by the security level of the object being written

   The rules above ensures a one way flow of information. In MAC some degree of centralization exists. Typically, a security policy administrator is responsible for maintaining the security levels of subjects or objects. The main disadvantage of MAC is its rigidity and centralized architecture makes it difficult to adapt it for distributed systems.

3. Role-Based Access Control (RBAC) has emerged in the past decade as the most widely discussed alternative to DAC and MAC. RBAC assigns permissions to well-defined abstractions called roles. Roles can be defined as a set of actions and responsibilities associated with a particular working activity [181]. Users then take on different roles in order to gain access to protected objects. RBAC allows a user to attain different permissions by switching to different roles in a given session [180].

RBAC makes the process of managing access rights easier by splitting the task of user authorizations into two parts namely, the assignment of access rights to a roles and the assignment of roles to users. This makes the assigning and revocation of access rights a convenient task. An advantage of RBAC is that it enables the creation a hierarchy of roles which makes it appealing to many highly structured systems. However, context is not fully considered in the activation, deactivation, and management of roles.

The access control policies discussed above are the most widely used policies in many security systems today. However, in the context of OSNs, where the social relationships play an important role in shaping the access policies, applying the traditional access control techniques to deal with the information sharing problem is not a trivial task. Still, there have been a number of studies developing different social networking based access control techniques, we review some of those papers next.

4.3.2   Access Control for Crowdsourcing Systems

Sharing of personal data is considered a major issue in OSNs and content sharing systems. Although some social networking sites like Facebook, Flickr, and Google Knol, have started to incorporate basic access control features into their sites, these controls are often limited and incomplete. Recently, a number of social networking based access control models have been developed to address the sharing problem.

The work in [203], [204] presents a social networking based access control scheme suitable for online information sharing. In the proposed approach, user identities are established through key pairs. Social relationship between users are represented as social attestations issued from one user to another and are used to gain access to friends' personal content. Access control lists (ACLs) are employed to define the access rights of the users based on the social relationships. To gain access to a particular object, a person must hold an attestation that satisfies the access control policies specified for the object.

Another access control model has been proposed in [64]. The authors introduce a rule-based access control mechanism for web-based social networks. The approach is based on the enforcement of complex policies expressed as constraints on the type, depth, and trust level of relationships existing between users. The model makes use of certificates to grant relationships authenticity. The authors propose the use of client-side enforcement of access control according to a rule-based approach, where a subject requesting to access an object must demonstrate that it has the rights for accessing the object.

In [207], Villegas et al. presented a personal data access control (PDAC) scheme for protecting personal data stored online. PDAC introduces a trusted distance measure that is computed based on the social network structural information and the experiential data of the users. Using the trusted distance, a data object owner defines three protection zones. PDAC uses a collaborative computing approach to map other users to the data protection zones defined by the owner of a data object. Based on the zone a user is mapped into, her requests to access the data objects of another user will be accepted, attested, or rejected. Attestation involves another round of evaluation by attesters designated by the owner of the data object.

The PLOG [106] is another access control scheme that is automatic, expressive and convenient. The authors are interested in exploring content based access control to be applied in social networking sites. Furthermore, Relationship-Based Access Control (ReBAC) was proposed in [88] to express the access control policies in terms of interpersonal relationships between users. ReBAC captures the contextual nature of relationships in OSNs. Relationships are articulated in contexts, and accesses are authorized also in contexts. Sharing of relationships among contexts are achieved in a rational manner through a context hierarchy. Authors also presents a policy language based on modal logic to express ReBAC policies. The language provides means for composing complex policies from simple ones.

In [164], [165], [166], we developed an equivalent set of users for information sharing instead of relying solely on topological constraints such as friends or friends of friends. This set would be given to a community of friends likely to have access to the shared information only if a quorum of them is given the information. We computed the equivalent sharing set as a connected component that surrounds a given user based on the communication patterns observed in the network. This set is referred as the α-myCommunity. For a given communication intensity threshold α, our algorithm gives a subgraph which is likely to contain the information that is being released by the user. If the user intends to actually restrict the spread of information to a subset smaller than the set found in this subgraph, then the risk of information leaking outside the user imposes a restriction that is given by the α value used in the subgraph computation.

4.3.3 Access Control Implementations in Crowdsourcing Systems

Most of the social networking services (e.g., Facebook, Google Knol, Flickr, etc.) incorporate basic access control features into their sites. The protection mechanisms implemented in most of the social networking sites are often very limited, allowing their users to set the confidentiality level for a given item as public, private, or accessible to friends. Certain social networking sites have incorporated variants of this simple protection scheme to provide their users with more flexibility in specifying the confidentiality levels.

For example, in addition to the basic access control options, Bebo (bebo.com) and Facebook (facebook.com) support friend-of-friend (2nd degree contacts) and "customized" (i.e., selected friend) access control options. Orkut (orkut.com) provides support for friends-of-friends. Myspace (myspace .com) and Google Knol (knol.google.com) only supports the basic options of public, private and 1st degree contacts. LinkedIn (linkedin.com) supports the option "my network," which is defined as the user's network of trusted professionals and includes 1st degree,

2nd degree, and 3rd degree connections in addition to members belonging to the user's LinkedIn groups. Flickr (flickr.com) supports public, private, and 1st degree connections (friends, family, or both).

It is important to note that all these simple access control schemes mentioned above, or variations of them, have several drawbacks. Many of these drawbacks stem from the following assumptions made by the simple schemes that are not always applicable.

1. **All friends are equal**. Access control schemes that use the hop distance to categorize friends assume that all friends at a particular hop distance are equal. While this makes access control simple, it lacks the flexibility of differentiating among the friends who are at the same distance when setting access options.
2. **Omniscient users**. Access control schemes that expect users to define access control lists to explicitly deny and allow accesses to data items assume that users are all knowing about their social neighborhood and able to make the appropriate protection decisions. With large and dynamic social neighborhoods (friends and friend-of-friends) such an assumption is not practical.
3. **No impact on friendships**. The simple hop-based protection scheme assumes that access control decisions do not impact the topology of the social network, which is often not valid. If Alice is unwilling to provide access to Trudy for data that she is already sharing with Bob, then it can indicate a lack of trust on Trudy.
4. **Friendships are static**. Because the nature of friendships can change from time to time, this can impact the owner's decision to grant or deny access for certain users to his data at certain times. To deal with this scenario, access control schemes utilizing only hop distance would require breaking and restoring the friendship links. Access control lists would require users to add and remove specific peers from each list; therefore it is a tedious task for users.

In addition to these issues, access control mechanisms used or developed for online social networking systems focus on protecting information resources within the system. In online social networks, information is often shared with other users for different purposes. Access controls techniques on social networks attempt to guarantee the protection of the information stored within the network, which means that once information is shared, those who gain access are able to use the content in anyway they want. Therefore in online social networks, there is a need for new access control schemes where the eventual information distribution is shaped by the initial release of the data items into the network.

4.4 Information Sharing Models

One of the important characteristics of OSNs is the private information space it provides for the users joining a network. After joining, users, at their discretion provide access to their friends using simple mechanisms provided by the OSN operators. Most OSN operators provide facilities to restrict access to subset of friends, friends,

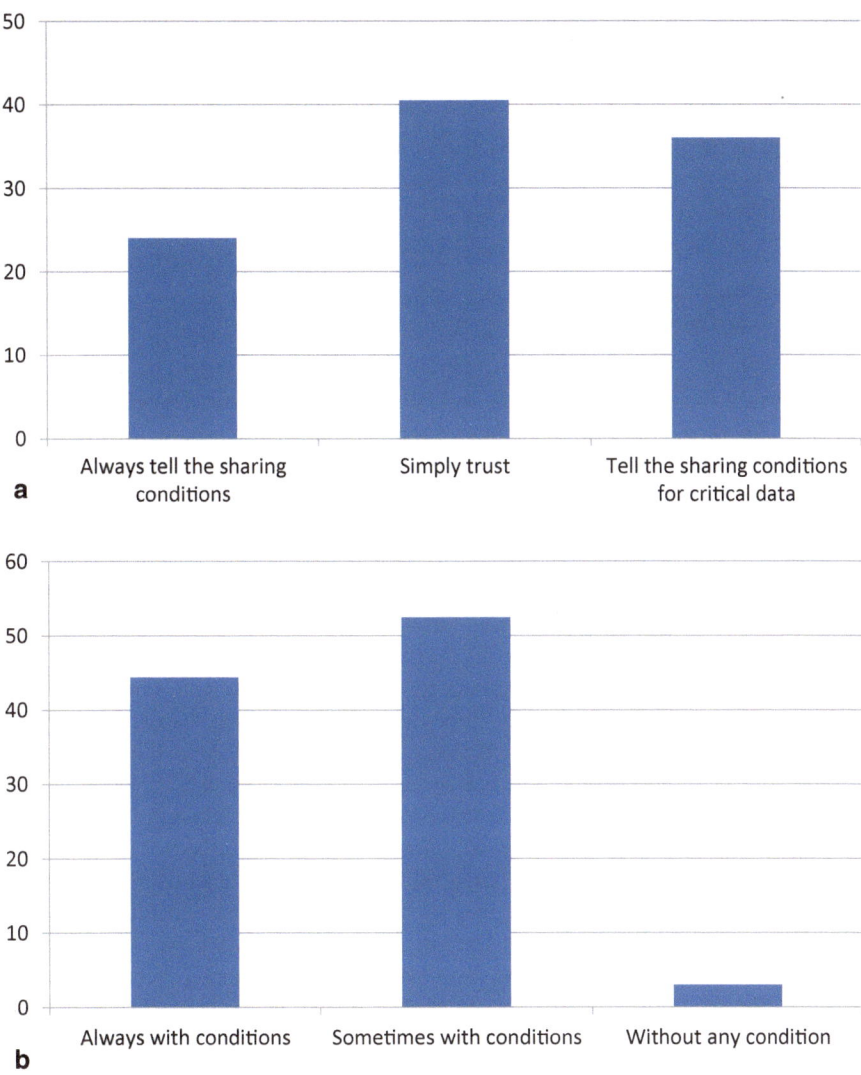

Fig. 4.1 Results of a survey on information sharing in OSNs **a** When friends share data with you, do you prefer them to state the conditions or simply trust you?, **b** Would you like to share information under specific conditions or under no conditions?

friend-of-friends, or public. These controls only deal with information release and expect the user to detect any misuse and modify the release conditions (for example, block an offending user from accessing data) [157], [225].

Information sharing in OSNs takes place without any formal usage restrictions or guidelines, which is an acceptable and preferable approach for OSN users as shown in the survey results shown in Fig. 4.1. This survey was conducted to find the value of information sharing on OSNs among McGill University students from various fields of study. Only 24 % of participants like explicit sharing conditions when they receive data from their friends whereas majority of the users prefer to attach specific constraints when they provide the information. This means in OSNs users expect to get access to friends' data objects without strict usage policy associated with them. They like to be "trusted" in handling the data that was shared with them by their friends. Users gaining access to a data object could leak it to an unauthorized users either unintentionally or maliciously.

4.4.1   Sharing Scenarios

In web-based social networks (e.g., Facebook, LinkedIn, Twitter, Google Buzz, etc.) sharing of personal information is most common. Information shared on these sites often consists of personal content or objects, including profile information, links, photos, videos, status messages, and comments displayed on users' personal pages. Generally, in any sharing situation the owner of the content expects the sharing conditions associated with the content to be honored. For the shared content, the value is related to the loss of benefit or amount of discomfort for the owner that results if the sharing conditions are violated. Therefore, for a high value object, the owner expects the other party (i.e., the user who has access to the object) to strictly adhere to the sharing conditions associated with the shared object. For example, if Alice shares a photo album with her friends and asks them not to share it with anyone else then, she anticipates that they will conform with her conditions. If one of Alice's friends (Bob) republishes her photos elsewhere then, her reaction to the situation will vary depending the value of the photos. From Alice's perspective the value of the photos depends on the level of discomfort (or loss of utility) caused from dishonoring her sharing conditions. If her loss is high, Alice can decide to blacklist Bob (i.e., stop sharing with him in the future) and also ask her friends to do the same (collaborative blacklisting). However, for Alice's friends, the value that they have for her photos can be different than her own value due to the fact that personal objects are valued *subjectively* by people in the social network. Depending on their valuation of the photos, Alice's friends might decide not to blacklist Bob. Thus, in an environment where the values of the objects are subjectively viewed, a collaborative blacklisting scheme might not necessarily work.

In addition to sharing personal information, social networks provide an ideal environment for collaboration among user communities for various reasons related to the exchange of knowledge and creation and sharing of common information goods. In such an environment, users are expected to cooperate with each other and remain active to derive utility from the available content. An example of such a case is a large collaborative community that is interested in developing and sharing software (e.g., open source software). The users in such a social network collaborate to design software or share their own developed software with each other. In such an environment, the users derive benefit from utilizing the content available in the network at different times. Thus, for the users, the utility gained is dependent on the continuous access to the content.

In a collaborative environment, objects are valued differently from the previous case (i.e., value of personal content). Here, the objects shared in a collaborative setting are *objectively* valued by the users. For example, if Alice spends a long time creating software that is popular within the user community, then the value of this software would be considered high by Alice as well as other users. The fact that the users agree on the value of Alice's object allow them to determine the loss in utility that results if the sharing conditions are violated. As a result, Alice can expect her friends to blacklist other users that violate the sharing conditions associated with her software. In addition, on social networks users working in a collaborative setting are often part of a close-knit community with strong ties; this makes the consequences of untrustworthy behavior by any of its users even greater. An example of such case is the close-knit community of Jewish diamond merchants [60]. In the diamond community, the consequences of cheating are severe; a merchant who is found to be untrustworthy is quickly known and expelled from the community. Therefore, a collaborative sanctioning (blacklisting) approach would be more acceptable among the members of a collaborative community.

4.4.2 Real Life Example

In November 2011, one Apple store employee named Crisp was fired for his critical Facebook posts. He posted negative messages about his employer on a Facebook friends' page assuming it was a private conversation. A coworker enrolled in the private group passed those messages to the Apple store manager [155] who fired Crisp. Although the messages were private, the communication was not protected because one of the friends decided to leak it. With all of the existing access control mechanisms in OSNs, Crisp was not able to prevent his posts from leaking to the manager. However, if Crisp knows about the risk of his information leaking out of the desired circle of friends, he can shape the information spread by not releasing the information to some of his friends. Because Crisp is only capable of controlling the release of information to his friends, he will use those decisions to maximize the information spread while the possibility of undesirable leakage is minimized. Better still, if Crisp could find the friends who should be blocked from getting access to the information, he could prevent it from flowing to his manager (his adversary) and he would not have gotten fired.

4.4.3 Information Sharing Protocols

Information sharing on the Internet uses many different service architectures, ranging from centralized client-server to fully distributed. The advent of large scale distributed applications has shown that a client-server model cannot address the reliability and scalability requirements. Experience with practical systems reveals that replicating the centralized servers would enhance fault tolerance but can lead to unwanted overheads and bottlenecks from synchronization mechanisms used for replica consistency. The peer-to-peer (P2P) crowdsourcing model represents a radically different and feasible alternative for many large scale applications in the Internet. With P2P systems, scalability is achieved by balancing the system load among the processes of the system. In this part, we discuss some popular crowdsourcing systems and trends for P2P information sharing.

Information sharing in distributed networks started with the development of the classic client-server model. In this model, users must continuously poll the server to obtain current data, resulting in multiple requests for the same data items. The alternatives to the client-server model for information sharing are based on data sharing models [70] where a set of publishers use channels to share information to a set of subscribers.

The P2P crowdsourcing sharing model provides a different alternative to the traditional client-server model. Applications built on the P2P model do not require their users to know about the location of the data, instead the system finds and returns the matching data, thus allowing the subscribers and publishers to focus on the content rather than its location. This is unlike client-server models where the applications require their users to know the location of the data to retrieve it. Thus, P2P models provide a more data-centric service structure compared to the traditional server-centric structure. Furthermore, the P2P distribution model exhibits three important characteristics of paramount importance for a distribution model: (a) self-organization: the network adapts itself to the changing topology (i.e., random departure and joining of peers does not break the system), (b) symmetric communication: peers act as both client and server, and (c) distributed control: there is no single point or centralized directory. Wide-spread use of the P2P distribution model in applications like SETI, Napster, and Gnutella indicate that there are many potential benefits to fully distributed peer-to-peer systems. These include providing more resilience and higher availability through wide-scale replication of content at large numbers of peers. Here, we discuss some recent approaches [174] to implement P2P distribution.

P2P crowdsourcing systems can be categorized as structured or unstructured. An unstructured P2P network is formed when the links between two nodes are established arbitrarily. Such networks can be easily constructed as a new node that wants to join the network can copy existing links of another node and then form its own links over time. However, in these networks, if a node wants to find a desired piece of data, the query has to be flooded through the network in order to find as many nodes as possible that hold the data. Some popular unstructured crowdsourcing networks are Gnutella, Bittorent, and Kazaa. Structured P2P networks maintain distributed hash tables (DHTs) at each node and hash functions are used to index and map each node and content in the network; a global protocol is then used to determine which node holds which content. Some popular structured P2P crowdsourcing networks are Pastry [175], Chord [191] and CAN [218].

Unstructured models are easily implementable, more scalable and highly fault tolerant compared to structured models. For instance, the unstructured models are more resilient to network breakdowns and have faster recovery time against churning than the structured models. These properties of the unstructured P2P model also make it the most suitable vehicle for epidemic style sharing techniques. Epidemic style information sharing has gained popularity as a potentially effective solution for sharing information in unstructured crowdsourcing systems, particularly P2P systems deployed on the Internet or ad hoc networks. This style mimics the spread of a contagious disease where nodes relay new information to other randomly selected nodes. Among epidemic protocols gossip is a suitable candidate for information sharing. The gossip protocol [76], [86] has been studied extensively in regard to information sharing in large scale dynamic networks where data and node changes are very frequent. Besides being used for sharing, gossiping has also been applied for aggregation [31], [117], network management [95] and data synchronization [150]. It was shown in [31], [117] that gossip protocols can perform as well as other epidemic protocols like random-walk [97]. However, the possibility of malicious behavior by nodes has been abstracted away in current uses of gossiping.

Rumor spreading [58] in large malice prone networks is similar to computer virus spreading. Several studies have been carried out [66], [69], [96], [118], [151] to model viral propagation using epidemiological models. In these models, a virus is considered as harmful spontaneously by all nodes in the system; thus a virus will be treated same way by all nodes. This approach is in absolute contrast with the trusted dissemination mechanism proposed in [146], [147], where a message considered as a rumor or spam by one node can be considered as fact by another node. This means that messages need to be treated differently at various points of the network without incurring extra overhead. Viral marketing [113] is another technique that allows sharing of information and its adoption over the network. This technique exploits social networks by encouraging their users to share product information. The process is analogous to the spread of self-replicating computer viruses. Though viral marketing started out with the simple word-of-mouth mechanism, it was quickly enhanced by network effects such as email services from Hotmail/Yahoo/Gmail attached advertisements to every email sent through them. Several sharing models [99], [102] have been used for modeling the adoption of products/brands by viral marketing. However, studies [131] have shown that viral marketing in general is not as epidemic as one might have hoped. It was suggested in [131] that viral advertising should analyze the topology and interests of the social network of their customers rather than using normative strategies for advertising.

4.4.4 Some Important Information Sharing systems

In this section, we discuss data dissemination in some crowdsourcing information sharing systems.

Usenet Usenet [178] is the oldest and most popular distributed message board and discussion crowdsourcing service available on the Internet where users read and post email-like messages called articles on to a number of distributed newsgroups.

Newsgroups are topical categories used for classifying articles. Usenet organizes articles into newsgroups and saves them on a news server. Individual users download and post messages to a single home server which is typically hosted by universities or ISPs. Each news server exchanges its local postings with other news servers it knows about. The news servers use flooding algorithms to propagate article copies throughout the network of participating servers. Users pull new articles from their home servers on demand.

Unlike typical P2P systems the connections among Usenet news servers remain fairly static and determined by the subscriptions. Moreover, Usenet does not consider trust of news servers and/or news articles to distinguish between true facts and rumors. However, techniques like collaborative filtering [122] are used to classify the articles in the most appropriate newsgroup. This classification process reduces a user's immediate visibility to a small set of articles but does not delete the unwanted articles from spreading within the news network.

Blogs, Wiki and Syndication Feeds Blogs or weblogs [13], [123] are the latest and most popular method for crowdsourcing of periodic articles by individuals. In blogs, only one user (owner of the blog) can post new articles while others can post follow up discussions with the blog owner having explicit control over how a subject is discussed on the site. Wiki [123] is a type of website that allows users to easily add, remove, or otherwise edit and change some available content, sometimes without the need for registration. Blog and Wiki articles are mostly shared via hot linking externally (e.g., emails) or by syndication. Syndicating techniques like RSS (really simple syndication) [105] and Atom [22] extract and aggregate headlines from the blog sites. Users pull the RSS/Atom aggregates and select the news item(s) that interest them most. These techniques do not involve trust for dissemination.

P2P File Sharing Networks/Content Delivery Networks P2P file sharing networks (PPFSNs) [40], [173] are well known for their ability to distribute content. These networks fetch content that is specified by users in some way (e.g., name of movie/song). Validity checking of the content requires the users to download the content in full and examine it. This means in PPFSNs it is not viable to compute trust on individual files. Instead, in these systems trust is computed solely for the servers to track the probability that they deliver valid content. More practical networks employ P2P distributed hash tables (DHTs) to compute and maintain the integrity/trust of servers and also of the content individually. However, these systems inadvertently give rise to other problems, e.g., attack on the DHTs to prevent correct data being returned [184]. Though research in this aspect has resulted in DHTs that use cryptographic techniques [93] to detect and ignore non-authentic data, they fail to address the attacks threatening the liveliness of the system by preventing nodes from finding each other. Several such attacks and their possible solutions are outlined in [184], however, they do not help to reduce the complexity of the systems. Another popular approach for sharing data efficiently on the Internet is via the content distribution networks (CDNs) [182]. CDNs take copies of content from a provider's server and cache it in servers spread throughout a network. However, CDNs are not concerned with the trustworthiness of the content they carry. Instead, they put explicit trust on the content providers to uphold quality.

Trust Driven Information Sharing   Trust driven information sharing proposed in [146], [147] is a framework for reliable sharing of information. The participating nodes (or news servers) mostly serve as clearing houses for news articles. Unlike Usenet servers, the nodes use trust-based evaluations to decide whether to accept or reject news originating from a particular node. Acceptance and publication of a new story (e.g., news article) is based on the story's social approval such as their popularity or evaluations/verifications provided by members/groups within the network. Further, connectivity among peers is not via subscriptions but based on social relations forged from story exchanges among the peers. The trusted sharing mechanism exploits the social activity among peers to find new peers and shun socially unacceptable peers.

Trusted sharing is a completely distributed infrastructure for trusted news sharing. The basic idea is a push-based sharing scheme that uses an inline filtering process such that the spread of untrustworthy messages are retarded while the spread of trustworthy messages are accelerated. However, it should be noted that trusted sharing involves disseminating only digests of the information being spread. The full content can be transferred by RSS/Atom approaches. In the trusted sharing mechanism, users (mostly) seek content that is previously unknown to them. Therefore, it is not possible to verify the validity by merely checking the integrity of the file. The primary approach is to verify the story by obtaining user opinions through a random sampling process. The authors showed that by using a combination of server and message trust, it is possible to present an unbiased collection of stories. This provides more flexibility as the news channel is still receptive of interesting stories from low trusted news sources.

Chapter 5
Integrity Management

5.1 Overview

Integrity management is the review or establishment of different mechanisms to en-sure the long-term integrity of artifacts in crowdsourcing systems. Crowdsourcing systems deal with two main challenges to preserve the integrity of their artifacts: assigning different capabilities to users, and evaluating users and their contribu-tions. Two extreme approaches have been introduced for maintaining integrity in crowdsourcing systems: Wikipedia style and Linux style [80]. With Wikipedia style, integrity emerges out of the crowd activity and can be less effective on sections of the encyclopedia that do not gain wide exposure. Incorrect or intentional bias can be introduced into the Wikipedia articles and remain there until subsequent readers flag the problems. On the other hand, Linux-style approaches tightly control the integrity by funnelling all changes through a single authority. This style resolves the inaccu-racy problem, yet increases the workload and creates a bottleneck due to having a single authority.

5.2 Integrity Management in Crowdsourcing Systems

Integrity in crowdsourcing information systems has been investigated by large num-ber of the research thrusts. They can be categorized into two groups. The first group investigates the trustworthiness of article whereas the second one is involved in the assessment of the integrity of articles.

The first group of study focuses on computing the trustworthiness of the content within the articles. The methods in this category offer a means for predicting the accuracy of some facts of an article. Cross [73] introduced an approach that calculates the trustworthiness throughout the life span of the content in an article and marks this by using different colors. Adler and de Alfaro calculated the reputation of the authors by using the survival time of their edits in the article [27]. Then, they analyzed exactly which text of an article was inserted by precisely which author. Finally, based on the reputation score of the respective authors, Adler and de Alfaro are able to compute the trustworthiness of each word [28]. Similar to Cross they illustrate the trustworthiness by using color-coding.

A. Ranj Bar, M. Maheswaran, *Confidentiality and Integrity in Crowdsourcing Systems,* 39
SpringerBriefs in Applied Sciences and Technology, DOI 10.1007/978-3-319-02717-3_5,
© The Author(s) 2014

The second group of study focuses on assessing the integrity of an article as a whole. A first work in this category was published in [134] introduced a correlation of the integrity of an article with the number of editors as well as the number of article revisions. The work in [135] defines three models for ranking articles according to their integrity level. The models are based on the length of the article, the total number of revisions and the reputation of the authors, which is measured by the total number of their previous edits. In [223], authors proposed to compute the integrity of a particular version of an article by feeding a Bayesian network with parameters such as reputation of the authors, number of words the authors have changed, and the quality score of the previous version. Furthermore, on the basis of a statistical comparison of a sample of Featured and Non-Featured articles in the English Wikipedia, authors of [193] constructed seven complex metrics and used a combination of them for integrity measurement. However, their work is completely based on the Wikipedia model as one of the widely used crowdsourcing systems.

5.2.1 Characterization of Integrity Management

In a crowdsourcing system where large number of users are involved in updating large number of documents, integrity needs to be enforced with due diligence. In most crowdsourcing repositories (e.g., Wikipedia) only one version of a document is retained. This means contention can arise due to conflicting opinions. To reduce the bias, crowdsourcing systems often include different sections for explaining different viewpoints. If an article is owned by a single user (e.g., Google-Knol), this problem does not arise. However, the article itself can be biased because the owner might be filtering out the opinions he disagrees with. This incurs the possibility of having multiple articles on a single topic, because users concerned about the bias of an existing article might start their own version to provide an alternative opinion.

In Table 5.1, we use different attributes to characterize twelve crowdsourcing information systems. The first attribute is the *type of identity* required by the site. Except Wikipedia, all other sites in this table require some form of login to create an article or edit contents that are already present in their sites. There are two types of identities: (i) real identity and (ii) pseudo identity. Sites such as Scholarpedia and Citizendium require real identity; i.e., the contributors have to register with their physical identifying information such as their curriculum vitae. Once the physical identity and the accompanying information of a user are verified, the user is given access to the site. Another form of identity that is more commonly required is the pseudo identity. In this case, users accumulate reputation on the pseudo identity and the privileges associated with a user are directly dependent on her reputation.

All sites, except Wikipedia, include ownership in their schemes to control the integrity of the generated articles. Therefore, editing capabilities can only be granted by the owner of an article (e.g., Scholarpedia and Citizendium) or by accumulating enough reputation (e.g., Stackoverflow). These systems try to encourage users to contribute more by providing incentives such as different capabilities within the system based on the level of their reputations. However, Squidoo or About sites provide some sort of remunerations to motivate users for contributing in their systems.

Table 5.1 Characteristics of some crowdsourcing websites

Name	Type of identity	Ownership	Editing capability	Need for approval	Rating	Scope of topics	Number of articles per topic	Remuneration
Wikipedia	Anonymous	No	Anyone	No	No	All	1	No
Knol	Google account	Yes	No, with owner's permission	No	Yes	All	Many	No
Scholarpedia	Real id	Yes	Yes, approved of author	Yes, expert	No	Scientific	1	No
Citizendium	Real id	Yes	Yes, before approval	Yes	No	All	1	No
Squidoo	Pseudo id	Yes	No	No	No	All	Many	Yes
Hubpages	Pseudo id	Yes	No	No	No	All	Many	Yes
Helium	Real id	Yes	No, owner	No	Yes	All	Many	Yes
Examiner	Real id	Yes	No	No	No	All	Many	Yes
Instructables	Pseudo id	Yes	No, contractor writers	Yes	No	Do-It-Yourself instructions	1	Yes
About	Real id	Yes	No	Yes, guides should accept it	No	All	1	Yes
DailyTech	Pseudo id	Yes	No, owner	No	Yes	Technology news	Many	No
Stackoverflow	Pseudo id	Yes	Yes	No	Yes	Programming	Many	No

Another important characteristic of these sites is the *number of articles per topic*. Depending on the nature of the site, we can have one or many articles for a specific subject. Wikipedia, Scholarpedia and Citizendium as online encyclopedias have only one article per topic. However, Google-knol, Squidoo, Hubpages, Helium, Examiner and DailyTech make it possible to have multiple articles when writers have different opinions on a specific subject. In addition, users can ask the same question multiple times in question answering sites such as Stackoverflow and get different answers.

5.2.2   Challenges

In this section, we discuss some of the challenges of creating crowdsourced information repositories. We categorize these issues as follows:

1. **Ownership of articles:** Suppose Alice is creating a document X. In a typical publishing scenario, she retains the ownership of document X for its lifetime. If anyone else wants to edit or modify it, they need to get Alice's permission. To this end, they have to submit their modifications to Alice and she has to decide which modification should be accepted or rejected. Obviously, such a publishing model is not very suitable for community-centric or large-scale collaborative publishing. In this model, Alice's ownership becomes a bottleneck because she must validate each and every modification. This is one of the reason that Wikipedia takes the ownership away from its core model. It attempts to provide integrity without requiring ownership.

2. **Accuracy of articles:** The involvement of large number of writers, often non-experts in Wikipedia-style websites, results in unreliable and unprofessional documents. Also, having a single version for each article may result in contentions that will need further resolution. The main problem here is the lack of authority and fact-checking. Someone should report the problem; otherwise, inaccurate information that is not obviously false may persist in Wikipedia for a long time before it is challenged [1].

3. **Duplication of effort:** Some crowdsourcing websites offer an option of having multiple articles on the same topic, each being written by a different author (owner). This can resolve the contention problem mentioned previously. However, in this model, a reader would have difficulty in searching and discerning the relevant articles from the irrelevant ones.

Majority of crowdsourcing websites employ the owner-centric model to solve the article accuracy problem experienced by Wikipedia. If a user creates an article, she is the owner of that document and also responsible for its credibility and future modifications. Since a user's reputation is highly dependent on the quality of articles created by the user, the user would pay more attention to the accuracy of her articles; hence, the quality of articles can be significantly improved. In the owner-centric model, a user can control the integrity by restricting the users who have access to the right to modify the content owned by the user.

5.2.3   Requirements

As mentioned earlier most crowdsourcing systems employ the owner-centric model which restricts the access to the modify rights for preserving the integrity of contents. Here, we discuss the requirements for all schemes used in restricting the access to the modify rights in owner-centric models [207]:

1. **Full control:** An access control scheme should provide the owner with full control on how she grants access to the modify rights to other users.
2. **Flexibility:** A crowdsourcing website can have various data and editing requirements. Therefore, the access control scheme should be flexible and capable of operating at different data granularities. This flexibility requirement has various aspects including: providing different levels of editing capabilities for different users, changing editing conditions for existing content (e.g., increasing its integrity level), and changing editing capabilities of a user (e.g., decreasing the trustworthiness of a user and consequently limiting her access to modify contents). While flexibility is essential in access control schemes, it can add significant overhead in terms of user effort required to setup and maintain an ownership model.
3. **Collaborative environment:** The user effort required by the access control scheme is certainly a major factor in its eventual acceptance in the user community. One way of retaining the flexibility and reducing the user effort required by the access control scheme is to enable collaborative decision making for each user. This can be done for a user–Alice–using one or a combination of the following methods: learning from Alice's past activities, learning from past activities of the community within a social neighborhood of Alice, and learning from the past activities of a set of users who are similar to Alice within her social neighborhood.
4. **Prediction of accessibility:** Another important requirement for an access control model is the ability to predict the accessibility of created contents. For instance, Alice would like to know the list of users who are able to modify the new article she is creating with a particular integrity setting. Such prediction of accessibility can be used to interactively shape the integrity settings for important data.

5.3   Access Control for Preserving Integrity

5.3.1   Major Access Control Schemes

Due to its importance, a number of solutions were proposed to the problem of preserving integrity in crowdsourcing systems based on different access control mechanisms. Here, we review some of the major techniques.

Task-Based Access Control   Task-Based Access Control (TBAC) [201] extends the traditional subject/object-based access control models by granting permissions in steps related to the progression of tasks. Permissions are associated with a protection state. This allows for type-based, instance, and usage-based access, and gives authorizations an expiration date. Weaknesses include having race conditions due to the "just-in-time" activation and deactivation of permissions, and the primitive nature of the specification of complex policies and management of authorization policies. A collaborative framework that uses TBAC in conjunction with RBAC has been presented in [115].

Team-Based Access Control   In most organizations, individuals are usually divided into teams that are used to associate a collaboration context with an activity. Team-Based Access Control (TMAC) [200] defines two parts of the collaboration context. User context allows identification of specific users that are taking on a role at any given moment, while the object context identifies specific objects that are required for collaboration. In addition to the strengths of RBAC, this model allows for fine-grained control on individual users in certain roles and on individual object instances. However, the TMAC model is not yet fully developed, and does not fully reflect the multidimensional nature of collaborative contexts such as organizational entities, workflow tasks, and components of groupwares [202].

Spatial Access Control   Spatial Access Control (SAC) [59] considers the environment of collaboration and uses it to hide explicit security mechanisms from users. It was developed primarily for collaborative virtual environments. One component of the model called a boundary divides the collaborative environment into regions and uses credentials to access these regions. The second component called an access graph specifies movement constraints in the collaboration space and manages the access requirements. However, this model does not provide fine-grained access control, and requires an application domain that can be divided into regions and boundaries.

Context-Aware Access Control   To assign different capabilities to users, crowd-sourcing systems often classify users into different groups, such as guests, regulars, editors, admins, and "dictators". Low-ranking users (e.g., guests, regulars) usually have few capabilities: answering questions, editing small part of artifacts, or flagging an incorrect data piece. On the other hand, high-ranking users (e.g., editors, admins) have wide variety of capabilities, from making small contributions to resolving controversial issues. This type of user classification is necessary to control the impact of a contribution. The potential impact of a contribution can be measured by considering how the contribution affects the crowdsourcing system. For instance, editing a sentence in a Wikipedia page can only affect that page, whereas revising a code in a software such as Linux can affect millions of users. Quantifying the potential impact of a contribution in complex crowdsourcing systems can be nontrivial [142], [143].

The main idea for evaluating users and their contributions is to detect spam and out of scope contributions and also differentiate low quality modifications from appropriate ones. Crowdsourcing systems utilize a combination of techniques to block and detect malicious users. First, crowdsourcing systems can block any malicious user by limiting who can make what kinds of contributions. For instance, Wikipedia blocks the IP address of a malicious user who has attempted to add irrelevant and inaccurate materials to an article multiple times.

Crowdsourcing systems can detect malicious users and contributions using two main approaches: content-based analysis and user-driven evaluations. Content-based analysis is an automatic method typically involves some tests. For instance, a system can ask users questions for which it already knows the answers, then use the answers of the users to compute their reliability scores [35], [143]. Many other schemes to compute users' reliability, trust, fame, or reputation have been proposed [116], [183]. User-driven approaches are manual techniques including monitoring the system by the users, distributing the monitoring workload among a set of trusted users, and enlisting ordinary users (e.g., flagging bad contributions).

Finally, crowdsourcing systems can deter malicious users with threats of "punishment." A common punishment is banning. A newer and more controversial form of punishment is "public shaming," where a user U judged malicious is publicly branded as a malicious or "crazy" user for the rest of the community (possibly without U's knowledge). For example, a chat room may allow users to rate other users. If the (hidden) score of a user U goes below a threshold, other users will only see a mechanically garbled version of U's comments, while U continues to see his or her comments exactly as written.

User-Driven Quality Evaluation Wikipedia uses user-driven approaches to differentiate between low and high quality articles. Wikipedia introduced the voting-based quality evaluations to tag articles as "Non-Featured Articles", "Good Articles" and "Featured Articles" [9]. Any user can nominate an article by listing it as a candidate for one of these categories. After the nomination of an article, it is flagged with a special tag. The article needs to meet a particular criteria based on the type of category to get the nominated category. Featured articles have the highest quality standard such as accuracy, completeness and should be well written. Good articles are also high quality articles, however, slight inconsistencies in the quality are tolerated (e.g., a lack of illustrations or small weaknesses in the writing style). Non-featured articles are the ones containing an unsuitable representation or a lack of relevance for Wikipedia. However, even this type of articles should meet a minimum standard of quality. The articles that are generally uncontroversial for deletion, such as those victimized by vandalism or other nonsense, are deleted quickly by using the speedy deletion procedure.

After the nomination of an article, the community decides whether or not the article belongs to a certain category via a voting process. The voting period and the voting rule depend on the kind of evaluation. For example, for a featured article, the voting period is 20 days and a two-third portion of the votes are necessary. After a successful election, the article will be added to the appropriate category by the addition of a special tag on top of the page [9].

Content-Based Analysis Wikipedia uses an anti-vandal detection mechanism called ClueBot [10] which utilizes machine learning techniques to detect user behavior and vandalism. ClueBot learns to detect vandalism automatically by examining a large list of edits pre-classified as either constructive or vandalism instead of using a predefined list of rules that a human generates. According to the Wikipedia page [10], ClueBot catches approximately 55 % of all vandalism correctly.In addition,

Wikipedia uses a software called XLinkBot [11] to deal with domains frequently misused by new and anonymous users. The XLinkBot allows established users to add links, while links added by others are reverted back.

Recent work have developed several approaches to quantify the integrity of artifacts in crowdsourcing systems. These approaches try to measure the quality of artifacts based on the length, the total number of revisions and the reputation of the editors [134], [135]. Blumenstock [52] demonstrated that the length of an artifact is the most accurate approach to distinguish high quality articles from low quality ones.

5.3.2   Social Integrity Management Scheme

In this section, we present the Social Integrity Management (SIM) scheme for preventing unauthorized modifications in online crowdsourcing systems. First, we introduce the different concepts in the design of SIM and then discuss the rationale that supports the choices made in developing the scheme.

In this scheme, users can view, vote, comment, edit and create (own) an article. We incorporate a number of social factors and ownership in our design to preserve the integrity of documents. It means that SIM utilizes user activities and the topological structure of the social graph to establish trust between the users. In an editing situation, we assume that an owner of the article is willing to accept modifications from a trustworthy editor. We apply these factors in SIM to categorize the users around the owner of the article.

Our approach for developing SIM is built on the following assumptions.

1. All users are part of a centrally maintained social network (e.g., Facebook).
2. Friendships among users on the social network are context independent (e.g., direct friends of Alice could include family members and colleagues at work).
3. The social network follows the best security practices in resisting whitewashing [75] and Sybil attacks [220], [222].

In SIM, we use an integrated namespace to facilitate discovery of articles for users. If a user wants to vote, comment, edit or publish an article, she should be logged into the system with her unique username. Having unique identifications helps to prevent multiple votes on one article from a specific user. More votes on one article indicate that it has higher level of acceptance among the users. Therefore, there are no anonymous commenting, editing or writing in the system.

Trust level The design of SIM leverages the structural properties of the social network and the activities of its users. In social networks, the structure of the relationships can be used to infer the degree of trust between its users [48, 98]. In such cases, trust is represented using the social distance between the users (hop-distance in terms of user relationships) in the social network. For instance, direct friends (1 hop friends) are considered more trusted than friend-of-friends (2 hop friends).

To quantify the trust between two users, we propose a *trust level* measure. Every user should classify her friends in different categories based on the level of trust between them. Friends can have three different level of trust: highly trusted, trusted and untrusted. Let us denote the trust level between two users x and y as $L_{trust}(x, y)$.

We assume that trust level is measured as a real number in [0, 1]. Therefore, the trust level for *highly trusted friends* can be represented as 1, for *trusted friends* as 0.5 and for *untrusted friends* as 0.

SIM scheme supports transitive trust between users. The trust level between users x and y will be the transitive trust for the smallest hop distance between them. The trust level between friends is automatically updated based on their activities in the system. These activities represent the history of the different editing activities that take place between the users. We will explain this process in detail in the voting section.

Set of trustable friends Based on the trust level, SIM categorizes friends around each user into different groups. Each user will have a set of trustable friends as follows:

$$T_x = \{y \in U | L_{trust}(x, y) \geq \beta\} \tag{5.1}$$

where, U is the set of users in the system, T_x is the set of trustable friends for user x, and β is the threshold. It means that if the level of trust between users x and y is greater than some threshold, then y will be considered as the trusted friend of x. In SIM, the trust level between two users (e.g., x and y) is asymmetric. User x can trust y to modify its articles (e.g., $L_{trust}(x, y) \geq \beta$); however, user y may not have enough trust on x in to give editing permissions to x.

Co-ownerships In SIM, any user who accepts to modify an article is considered as the co-owner of that article with the same privileges. Therefore, the set of co-owners for a particular article includes the creator and all other users who have edited the article so far. The creator of an article is considered the originator and all co-owners are responsible for the credibility and future modifications of the article. We incorporate co-ownerships in our model to prevent the article accuracy problem in existing systems such as Wikipedia. Since there is no anonymous modification, an article represents the opinion of the authors who put their reputation on the line. The users pay more attention to the accuracy of their articles; hence, the quality of articles can be high.

In addition, co-ownership limits the editing capabilities to only trustable users. In open systems such as Wikipedia, users have to watch their articles and revert back inaccurate and unreliable modifications. This results in large efforts for users to keep the integrity of their articles. Having co-ownerships provide incentive for other users to participate in the modification and evolution of an article. SIM scheme supports three different mechanisms for finding the set of trustable editors for a particular article, limited, semi-limited and open access.

- *Limited*: In this case, the editing access will be limited to users who are considered trustable by all co-owners of the article.
- *Semi-Limited*: a user will be considered trustable if at least q percent of the co-owners consider her trustable.
- *Open Access*: a user will have editing privileges if at least one of the co-owner considers her as a trustable editor.

Limited number of versions Co-ownerships reduce the control the creator has on the article because no agreement is necessary for co-owners to modify the article. In addition, depending on who accepts to edit an article, we can have different sets of co-owners for one particular article. This results in having different versions for specific topic with different integrity levels. Number of versions for a particular topic can be either constant k (e.g., $k = 3, 4, ...$) or p percent of the total number of documents.

Voting In SIM, there is a voting process to select the most popular articles for each specific topic. To obtain the highest number of votes, owners of an article push their article to their trustable friends to get their feedbacks and also involve them in the evolution of their article. When a trustable friend accepts to modify an article, she is considered as one of the co-owners. Therefore, the new co-owner as well as previous co-owners try to improve the article to get the highest number of votes. The more trustable friends get involved, the higher the chance that the article becomes one of the most popular ones.

After an article gets sufficient amount of contributions, any of the co-owners can nominate the article by listing it as a candidate for the most popular article in a particular topic. After the nomination of an article, it is flagged with a special tag called nominee and the community decides whether or not the article belongs to the most popular ones via a voting process. The voting process includes a voting period which the co-owners have time to advertise their article to get large number of votes. After a successful election, the SIM scheme presents the most popular articles based on the number of votes with the list of all co-owners.

Accordingly, the trust level between friends is automatically updated based on their activities in the system. These activities represent the history of the different editing activities that take place between the users. When a trustable friend, user x, accepts to edit a particular article automatically the trust level between her and all the co-owners will be updated based on the result of the voting process. If the article becomes one of the most popular ones, the trust levels between user x and all the co-owners will be increased by a constant value α. This means that the initial trust level values that the co-owners assigned to her friend at the beginning of the friendship were accurate and now they can even have higher trust on her trustable friend. On the other hand, if the article cannot pass the voting process, the SIM scheme automatically decreases the trust level between the co-owners and the user x by a constant value γ.

5.4 Trust and Reputation

As the Internet evolved and began to display the behaviors of the mainstream community in which it operates, trust and reputation systems gained significant importance for supporting services on the Internet. Trust in human life is a multi-disciplinary concept and similarly the related research literature for trust in Internet and distributed systems also presents several definitions [140, 196]. In this section, we focus on the concept of trust and reputation in online crowdsourcing systems.

5.4.1 Definitions

Simply put, trust is a judgment of questionable utility that helps to cope with uncertain situations created by others. One of the oldest and widely accepted definitions of trust is in [78] which stated that: "trusting behavior occurs when an individual confronts an ambiguous path, the result of which can lead to an event perceived to be beneficial or harmful for the individual and the occurrence of the result is contingent on the actions of another individual; if the individual chooses to take an ambiguous path with such properties, he makes a trusting choice; if not he makes a distrustful choice".

The definition shows the basic structure of a trusting choice; illustrating the concept of dependence on trusted parties. Trust is required for deciding about those actions which are difficult to control and monitor individually. A trusting choice manifests an individual's extent of belief in others with the anticipation that the decision can result in good or bad consequences. In other words, trust is considered to be a subjective quality which individuals place upon one another. In [94], the authors further refine the above definition to describe trust as a particular level of the subjective probability with which an agent can rely on the actions of another agent, both before he can monitor such actions and in a context in which it affects his own actions. In addition to the subjective nature of trust as defined in [78], the authors in [94] imply that the level of trust depends on how one's own actions are affected by other's action. This definition points out the fact that a trusted choice may lead to a change of behavior towards the trusted parties and either strengthen or reduce future cooperations with them.

Another definition of trust was published in [101] which reports trust as the firm belief in the competence of an entity to act dependably, securely, and reliably within a specified context. This definition, though rather vague, explicitly and implicitly incorporates the various notions of trust: (i) dependence on trusted parties, (ii) reliability of the trusted party, and (iii) the risk involved in trusting others as it may result in beneficial or harmful consequences.

Based on the above definitions, it can be gathered that trust is linked to the confidence placed on someone or something, whatever it is the desirable outcome is contingent upon. Trust implies some degree of uncertainty while attaching some degree of hopefulness or optimism about the outcome. Thus, some specific properties of which trust can be attributed to the topic at hand can be summarized as follows [156]:

- Trust is subjective. The subjective nature of trust is one of the most challenging properties of building a trust relationship, because parameters used in each trust-related transaction varies widely depending on the concerned individuals.
- Trust is context specific. Trust is multi-dimensional and varies according to the purpose of the system. For instance, Alice can trust Bob completely when it comes to repairing cars but Alice may not trust Bob for repairing computers.
- Trust is only conditionally transitive. Alice can trust Bob who in turn can trust Cathy. But this does not require Alice to trust Cathy. However, Alice can trust Cathy if certain application specific conditions are met.

- Trust is measurable. It is possible to measure the value of trust, which allows comparison of trustworthiness. The ability to measure also helps competition, which is useful to break the monopoly of trusted parties.
- Trust is dynamic. Trust reasoning is dynamic and non-monotonic. Values of trust may increase or decrease with time and context.

Related to trust is the concept of reputation. In a generic sense, reputation is what is generally said or believed about an individual's character or standing [112]. This definition is further elaborated in [25] to link reputation to the concept of trust. It describes reputation as an expectation about an individual's behavior based on information about or observations of its past behavior. In other words, reputation is considered as a collective measure for trustworthiness and a form of social control.

In almost all definitions of reputation, it is considered as a propagated notion, e.g., passed via agents or via "word-of-mouth." In online communities, reputation is approached by the recommendation operation and implied as a quantity derived from the underlying social network which is globally visible to members in the network. In a centralized system, a single agent collects the recommendation information for other entities in the system and infers reputation estimates for them.

eBay's feedback system [170] is a good example of a centralized reputation system. Centralized reputation systems work based on the assumption that people trust the reputation information presented to them, which is true for eBay. However, in a distributed crowdsourcing approach, there is no large, recognizable organization. Instead, distributed crowdsourcing systems heavily rely on the "word-of-mouth" mechanism where each entity will share with the rest its opinion about others in the system. Such a mechanism, however, relies on a social network to gather and disseminate reputation information.

Reputation computation involves different kinds of information. These can largely be classified as positive reputation, negative reputation and a combination of both. Positive reputation typically provides positive feedback about agents, e.g., in a distributed file sharing system, positive reputation for a file server is given by the number of successful downloads from the server. Along the same lines, negative reputation generates negative feedback, e.g., number of failed or unsuccessful downloads from a file server. Both types of reputation information are, however, incomplete without each other. Simply relying on one type of reputation information results in bias. For example, in an information sharing system, relying on positive feedback about a peer will result in ignoring the recent malicious actions by the so called good peer. Simply depending on negative feedbacks is also harmful since in a negative reputation mechanism, entities are generally assumed to be good unless proven guilty. So, peers may end up trusting a malicious peers due to lack of complaints about the malicious peer.

A non-bias reputation system, specifically for distributed systems, has very specific properties [169]. These are as follows:

- Longevity of agents. Entities should be long lived so that for every interaction there is always an expectation of future interaction. Longevity ensures that agents may not change identity to erase past behavior.

- Feedbacks for current interactions can be captured and distributed. This ensures that agents provide feedback to others in the system. However, this depends on the willingness of the participants.
- Feedbacks about past interactions must guide decisions about current interactions. This ensures an accurate response model for the reputation system.
- Feedback should be context specific. This ensures that agents are not penalized for their behavior for unrelated interactions.

5.4.2   Representative Trust Management Systems

Commercial and Live Models   Slashdot's and eBay's reputation systems are good examples of commercially used reputation based trust management systems. The Slashdot reputation system [126] uses a two-level approach to compute ranks for article comments. In this approach, the lower layer computes an initial score for newly posted comments which are then reviewed and sanctioned (or rejected) by the members of the upper layer. The reviewers are trusted members of the Slashdot user community. Users obtaining high scores for their comments are considered as trusted members and earn the privilege of posting new articles.

eBay's reputation system [170] is a centralized system where buyers and sellers rate each other for each transaction. The central system collects the ratings and computes reputation for each member. This model has suffered from many ballot stuffing (i.e., repeated ratings using fake transactions) attacks. Another (almost) centralized reputation model was introduced by Advogato for online software development communities [12]. It is based on a trust flow model, where the flows originate from few highly trusted nodes. A trust flow is implemented by a node providing a referral to another node with a referral always flowing downstream. This makes the model highly dependent on the integrity of the upstream nodes which acts as trust managers for downstream nodes.

Academic and Research Models   PolicyMaker [50] is an example of a credential and policy based trust management system. In PolicyMaker, a peer grants access to its services when it can verify that the requesters' credentials satisfy the policies needed to access its services. PolicyMaker provides specific language syntaxes for describing policies, credentials and complex relationships. Security policies and credentials are defined in terms of predicates called filters that are associated with public keys. The trust management system in PolicyMaker simply verifies the trust relations to allow accesses. For example, the filters accept or reject actions based on what the bearers of the corresponding public keys are trusted to do. Credentials are interpreted with respect to the policy statements and can return a positive or negative response. In distributed systems, the PolicyMaker service is used as a remote database query service via a set of APIs to determine the policies while the application using PolicyMaker is responsible for performing the key verification functions. Other trust management systems follow similar principles as Policymaker [50]. The only difference being in these systems the key verification is done by the trust management engine and not left to be done by the applications.

A quick study of the literature shows that a large number of trust management systems are based on reputation. In a framework called EigenTrust [114], peers rank their transactional experiences with others to form local reputation estimates about the transactors.The local estimates are then exchanged (via recommendations) with other known peers to form global aggregates. The global aggregates are used as trust estimations for making system-wide transacting decisions. A similar approach is followed in [74], [79], [104] except that local reputations are verified against findings of trusted third parties to form global trust estimates. In contrast to these models, a hierarchical trust management framework in [148] showed that by splitting local and global trust computation along different layers and using explicit trust managers, it is possible to achieve much higher prediction accuracy.

The XRep protocol [74], belonging to the reputation-based trust management category, evaluates the quality of a service or resource provided by a peer besides modeling the behavior of the peer. It uses a distributed polling algorithm for sharing reputation information with other peers in the system. In XRep, each peer in the application termed as a servant maintains information of its own experience on resources and other peers. This information is shared with other servants upon request. Servants store the information in two repositories: (i) a resource repository which associates a unique resource ID with a binary reputation value, and (ii) a servant repository which associates with each unique servant ID the number of successful and unsuccessful downloads.

The distributed polling algorithm for sharing the reputation consists of resource searching, resource selection and vote polling, vote evaluation, best servant check, and resource downloading. Resource searching, similar to searches in Gnutella, involves a servant broadcasting to all its neighbors a query message containing the search keywords. When a servant receives a query message, it responds with a QueryHit message. Next, upon receiving QueryHit messages, the originator selects the best matching resource among all possible resources offered. At this point, the originator polls other peers using a Poll message to enquire their opinion about the resource or the servant offering the resource. Upon receiving a poll message, a peer responds by communicating its votes on the resource and the servants using a Poll-Reply message for this purpose. These messages help identify reliable resources from unreliable ones, and trustworthy servants from fraudulent ones. In the third phase, the originator collects a set of votes on the queried resources and their corresponding servants. After receiving the votes, the peer begins a detailed checking process which includes verification of the authenticity of the PollReply messages. At the end of the checking process, based on the trust votes received, the peer may decide to download a particular resource. However, since multiple servants may be offering the same resource, the peer still needs to select a reliable servant. This is done in the fourth phase where the servant with the best reputation is contacted to check whether it can export the resource. Upon receiving a reply from the servant, the originator finally contacts the chosen servant and requests the resource. It also updates its repositories with its opinion on the downloaded resource and the servant who offered it.

Another reputation-based trust management system is described in NICE [129], which is a graph-based model for evaluating recommendations. The NICE algorithm uses a directed graph called trust graph, where each vertex corresponds to a peer in the system. A directed edge from peer A to peer B exists if and only if B holds a cookie signed by A, which implies that at least one transaction had occurred between them. The edge value signifies the extent of trust that A has in B and depends on the set of A's cookies held by B. If A and B were never involved in a transaction and A still wants to compute B's trust, it can infer a trust value for B by using directed paths ending at B. Two trust inference mechanisms based on such a trust graph are described in the NICE approach: the strongest path mechanism and the weighted sum of strongest disjoint paths mechanism. In strongest path mechanism, strength of a path is computed either as the minimum valued edge along the path or the product of all edges along the path. In weighted sum of strongest disjoint paths, peer A can compute a trust value for B by computing the weighted sum of the strength of all of the strongest disjoint paths.

An important contribution of the NICE approach is the ability of good peers to form groups and isolate malicious peers. In order to form such groups efficiently, peers maintain a preference list of potentially trustworthy peers. The list is constructed based on previous interactions and observations. This ability to form robust cooperative groups, along with the incentive to store cookies, improves the reliability of the system. NICE employs the use of both positive and negative cookies to achieve a more robust reputation scheme. NICE works in a purely decentralized fashion and each peer stores and controls data that benefits itself. Therefore, storage costs even in the worst case are limited by the number of interactions with other peers. Further, to improve the efficiency of the cookie-search mechanism and to limit bandwidth costs, NICE employs a probabilistic flooding-based search mechanism.

Another very successful approach used for reputation-based trust management is called collaborative filtering (CF). In collaborative filtering, user ratings for objects, like news articles/products, are used to determine which users' ratings are most similar to each other and predict how well users will like news articles/products from users giving similar ratings. Primary usage of CF is to predict new items to users based on the underlying assumption that those who agreed in the past tend to agree again in the future. A centralized approach for making CF predictions is described in [30], [122]. In [122], a CF mechanism is applied to Usenet news articles such that articles can be disseminated to users interested in them. The CF mechanism applies the Pearson's correlation formula [122] on the ratings users provide for the news articles to predict. The assumption being, if users are likeminded, they are more likely to be interested in same or similar news articles. However, there are few disadvantages of this model. Firstly, the user opinions can be biased and the system does not provide a way to correct the bias. Second, sufficient data is required to make accurate predictions. And lastly, a CF system is not very scalable, since the CF computations are centralized; increasing the number of users or new articles exponentially increases the computation complexities. To overcome these disadvantages, in [30], a graph-theoretic approach was proposed to do collaborative filtering. The basic idea was to create and maintain a directed graph whose nodes are the users and whose directed edges correspond to the predictions.

Social network-based trusted systems have gained popularity in recent times. The Regret system [176] includes the social relations of peers and their opinions in its reputation model. Regret assumes that the overall reputation of a peer is an aggregation of different pieces of information: individual, social, and ontological which form the three dimensions for reputation. Combination of these three dimensions yields a single value for reputation. When a peer depends only on its direct interaction with other members in the system to evaluate reputation, the peer uses the individual dimension. If the peer also uses information about another peer provided by other members of the society it uses the social dimension. The social dimension relies on group relations and a peer inherits the reputation of the group it belongs to. As a result, the group and relational information can be used to attain an initial understanding about the behavior of the peer when direct information is unavailable. Thus, there are three sources of information that help peer A decides the reputation of a peer B: (i) the individual dimension between A and B, (ii) the information that A's group has about B called the Witness reputation, and (iii) the information that A's group has about B's group called the Neighborhood reputation. Furthermore, Regret believes reputation to be multi-faceted. For example, the reputation of being a good flying company summarizes the reputation of having good planes, the reputation of never losing luggage, and the reputation of serving good food. In turn, each of these reputations may summarize the reputations of other dependent factors. The different types of reputation and how they are combined to obtain new types of reputation is defined by the ontological dimension. Clearly, because reputation is subjective, each peer typically has a different ontological structure to combine reputations and has a different way of weighing the reputations when they are combined.

The Regret trust model is a decentralized system where each peer maintains local control over its own data. It also considers group reputation and the ontological dimensions while computing reputation. This increases Regret's flexibility and reliability. However, in Regret, a peer does not cross group-boundaries to inquire peers from other groups about the reputation of a peer. Clearly, if this were to be implemented, the reputation model would become quite complex and would require increased communication between peers. In addition, while the existing model is simple, each peer assumes an implicit trust in other peers belonging to the same group, thus, exposing itself to malicious activity within its own group. Regret expresses trust using both positive and negative levels of reputation. Since each peer stores group information in addition to peer information, additional storage space is required. The main shortcomings of the Regret model are the lack of credential verification, techniques to protect users' anonymity, and fault tolerance mechanisms. Upon detection of fraudulent actions, affected peers can modify not only the reputation value of the malicious peer but also that of the witnesses that recommended the peer. These changed values will forewarn other peers in the future. In addition to using this technique, a peer can combine opinions of multiple witnesses to detect misrepresentation. In Regret, collusion can be prevented by using the social reputation mechanism as long as the number of good peers is sufficiently greater than the number of malicious peers. New peers that join the system start with zero reputation but quickly build up their reputation through successful interactions.

NodeRanking [163] is another social network-based trust management system. The goal behind reputation systems like NodeRanking is to remove dependence upon the feedback received from other users; instead, explore other ways to determine reputation. NodeRanking views the system as a social network where each peer has a social standing or position in the community. The position of a peer within the community can be used to infer properties about the peer's degree of expertise or reputation. Peers who are experts are well-known and can be easily identified as the highly connected nodes in the social network graph. Other peers can use this information directly instead of having to resort to explicit ratings issued by each peer. The NodeRanking algorithm helps create a ranking of reputation values of a community of peers by using the corresponding social network. The reputation value for a peer is based on the concept that each node on the graph has an associated degree of authority that can be seen as an "importance" measure. During bootstrapping, the system assumes that all nodes have the same authority. The NodeRanking algorithm is then executed to calculate the authority values of all peers in the system. The social network can be considered as a directed graph where each edge has a direction. Edges that start from a node are called its out-edges and the nodes that they connect to are called its out-nodes. Similarly, edges that come into a node are called its in-edges and the nodes that they start from are called its in-nodes.

The core idea behind NodeRanking is that each node has an authority and a part of this authority is propagated to its out-nodes through its out-edges. Thus, the authority of a node depends on the authority of its in-nodes. The authority measure of a node is calculated as a function of the total measure of authority present in the network and the authority of the nodes pointing to it. Nodes that are not pointed to by any other node are assigned a positive default authority value. The resultant authority values obtained after executing the NodeRanking algorithm are used to infer the reputation of the peers in the community. The principal shortcoming of NodeRanking is that it is centralized. While the NodeRanking algorithm does not require each peer to know about the rest of the system, the results from each peer are returned to a centralized node in order to construct the social network graph. This centralized node is then queried for reputation information by the peers.

NodeRanking inherits the disadvantages of any centralized scheme: single-point-of-failure and issues of scalability. Moreover, there is no authentication of the communication between the centralized node and the peers and no mechanism to protect the privacy of peers. NodeRanking lacks a scheme for preventing a malicious peer from misrepresenting authority or to protect against a group of colluding malicious peers that may point to other peers in their own clique to increase their authority.

5.5 Trust Model for Web-based Crowdsourcing Systems

The term trust management was first introduced in [51] as a unified approach for interpreting policies, credentials, and relationships which allow direct authorization of security-critical actions. However, complexities of recent crowdsourcing systems have resulted in a much broader definition that is no longer limited only to authorizations. In [100], trust management is described as "the activity of collecting, encoding, analyzing, and presenting evidence relating to competence, honesty, security or dependability with the purpose of making assessments and decisions regarding trust relationships." In general, trust management systems can be classified into three categories: (a) credential and policy based trust management, (b) reputation based trust management, and (c) social network based trust management. This categorization is based on the approach adopted to establish and evaluate trustworthy relationships among peers. This section discusses the categories using representative application models.

5.5.1   Credential and Policy-Based Trust Management

In this approach, credential verification is used to establish trust relations among the peers. The primary goal is to perform access control based on a set of credentials and a set of policies. The concept of trust management is, thus, limited to verification of credentials and restricting access according to application defined policies. Access to a service (or resource) is granted only if the service requestor's identity can be established and checked against the policies allowing or restricting access to the service.

The policy based approach was developed within the context of structured organizational environments. Hence, they mostly assume very specific sources for trust management such as certification authorities (CAs). For deploying this approach in open distributed systems, it is necessary to develop policy languages and engines for specifying and reasoning on rules to determine whether or not an unknown user can be trusted. However, besides access control, research has shown that it is also possible to formalize trust and risk assessment using logical formula [188].

This approach is best suited for access conditions which eventually yield a boolean decision, i.e., grant or deny access. Systems enforcing policy based trust use languages with well-defined semantics and make decisions based on attributes which can be verified by some authority, e.g., certificates obtained from CAs, date of birth obtained from governmental bodies. Policy based trust management is mostly intended for protecting systems whose behavior can be easily changed but where nature of information used for the authorization process is exact. This puts policy based approaches in direct contrast to reputation and social network based trust management systems which are ideally developed for unstructured use of communities where behavior of users do not only change randomly but the available data for authorization may also be limited.

5.5.2   Reputation-Based Trust Management

A reputation system attempts to compute a global estimate of a node's (or entity's) trust. They provide mechanisms by which a node requesting a service (or resource) may evaluate its trust in the reliability of provider nodes and the services being provided by them. The trust value assigned to a trust relationship is a function of the provider node's global reputation and the requesting node's perception of the provider node [114], [126], [170].

Trust recommendation models combine recommendation protocols with a trust evaluation system [24], [168]. This approach allows the system to be decentralized, generalize trust computation schemes, and use community feedback (i.e., recommendations). Decentralization allows peers the autonomy to create their own trust policies. Each node in the network works independently without having to update the rest of the network about their policies. Generalized trust computation schemes allow applications to use different metrics for trust evaluations such that trust can be expressed with respect to some context. Community feedbacks help to update the whole network about the behaviors of an individual node. This way, uncertainties arising due to unknown nodes can be resolved - nodes query their neighbors to find out about other nodes they have not transacted with personally. A recommendation is some reputation information that identifies the communicating node's experience with another node and is based on some metric plus context. In addition, in this approach, trust can be expressed as direct between two transacting nodes or conditionally transitive.

5.5.3   Social Network-Based Trust Management

A social network is basically a social structure of nodes tied by some relations. In most existing social networks, the network consists of real people and the relations among them are depicted by their like mindedness or similarity of their interests. A social network based trust management system analyzes the social structure of the network to form conclusions about the nodes. For example, a common wisdom in social networks is the more edges, the better, since within a social structure a popular or trusted person is known to have relations with a large number of nodes in the network. So considering the social graph, a simple count of a node's incoming edges will give an indication of the node's trust estimate within the network, where the incoming edges represent the number of other nodes willing to trust it. In most practical situations, however, a simple count of edges is not always sufficient. What really matters is where those edges lead to and how they connect to the otherwise unconnected nodes. More complex models are, thus, built by including more information about the edges.

Chapter 6
Case Study: Integrity of Wikipedia Articles

6.1 Overview

In this chapter, we investigate the integrity of Wikipedia articles as an example of integrity management in crowdsourcing systems. We first challenge its integrity by performing some experiments. Afterwards, we analyze the dump datasets from its website to find the reasons behind high integrity for few articles and low integrity for majority of them.

6.2 Challenging Wikipedia's Integrity

To meet its integrity objectives, Wikipedia encourages contributors by facilitating easy article creation and update. The integrity of a user's contributions are checked and flagged by subsequent readers. For highly trafficked articles, this model of integrity enforcement works very well. However, when integrity emerges out of the crowd activity, it can be less effective on sections of the content that do not gain wide exposure. Incorrect or intentional bias can be introduced into the Wikipedia articles and remain there until subsequent readers flag the problems.

In Wikipedia, there are two different modes for user contribution: anonymous and pseudonymous. In the anonymous mode, writers will not have full privileges and Wikipedia keeps their IP addresses in the history of the page. With pseudo identities, users accumulate reputation and the privileges associated with the users are directly dependent on the corresponding reputations.

To examine how Wikipedia preserves the integrity of its pages, we conducted some experiments by creating new pages containing false information and by modifying existing pages by adding non-related sentences and URL links.

6.2.1 Creating a New Page

We tried to create a new page with invalid content. Every new page proposal by an anonymous or new user goes through a validation by one of the Wikipedia's editors.

A. Ranj Bar, M. Maheswaran, *Confidentiality and Integrity in Crowdsourcing Systems,* 59
SpringerBriefs in Applied Sciences and Technology, DOI 10.1007/978-3-319-02717-3_6,
© The Author(s) 2014

The editor will check whether the page already exists, has commercial or offensive content or includes invalid information. After validation, the proposed page will be added to Wikipedia. The page we created could not pass the validation phase as we were anonymous or new user and had invalid content.

6.2.2 Modifying an Existing Page

Next, we attempted to modify an existing page by adding random sentences. New or anonymous user's modifications to existing pages go through an automatic validation process. Wikipedia uses an anti-vandal detection mechanism called ClueBot [10], which utilizes machine learning techniques to detect users behavior and their potential for vandalism. Our modifications were flagged by this software as invalid and never applied to Wikipedia pages. To mislead the ClueBot validation, we tried to add sentences with incorrect information containing words related to the page's topic. In this case, the ClueBot failed to detect our modifications.

6.2.3 Adding Random URL Links

In this experiment, we tried to add random, non-related URL links to popular and unpopular pages. We realized that all added links from new or anonymous users go through a validation phase similar to the previous experiment. Wikipedia uses a software called XLinkBot [11] to deal with domains frequently misused by new or anonymous users. The random URL links, introduced by us, were detected and removed by XLinkBot. To mislead the XLinkBot validation, we attempted to modify references in some pages by copying existing URL links from other Wikipedia pages. Here, we had two different scenarios where XLinkBot becomes confused. In the first scenario, the topic of the two pages are completely different and non-related. Since the copied link has no relation to the page's topic, the XLinkBot is still able to detect this modification and remove the link. In the second scenario, we considered two pages which are related to each other, i.e., one page has linked to the other one. In this case, XLinkBot failed to detect our modifications.

After running these three experiments, we can conclude that the integrity of the page depends on the readers who are responsible to find unrelated references and remove them by rolling back the page. However, when Wikipedia readers have less interest in some articles, their integrity may not be heavily scrutinized as pages with heavy readership.

6.3 Analyzing Wikipedia's Integrity

To solve the integrity problem, Wikipedia has developed various approaches for evaluating its articles. Wikipedia provides a user-driven approach where users can vote for articles to be marked as *"Featured Articles," "Good Articles,"* or *"Non-*

Featured Articles." Here, we aim at providing a better understanding of how an article becomes a featured article while others remain at low quality.

In this integrity analysis, we used the Wikipedia dump dataset for a period of ten years from 2001 to 2011. The dataset includes XML files containing the source texts of all pages with their complete edit history. The edit history contains the usernames or the IP addresses of the editors and the modification times. Because the size of the dataset is very large, Wikipedia divided the dataset into many files; each one contains information for around one thousand pages. Here, we only considered 100 good and featured articles as the sample for high quality articles and 100 non-featured articles as the test-case for low quality articles. This is a reasonable amount of data given that there are only 3,783 featured articles in Wikipedia which is around one of every 1,100 articles. Also, note that extracting the information about these types of articles is an extremely time consuming procedure. In addition, there are many low quality articles with very few contributions. Here, we only consider an article as part of our test-case if the article has more than 100 modifications in its edit history.

6.3.1 Contributions and Reverted Back Modifications

The number of contributions and the number of reverted back modifications for both low and high quality articles are shown in Fig. 6.1. For low quality articles, the number of reverted back modifications is similar or larger than the number of accepted contributions. It seems that Wikipedia community for low quality articles could not agree on the content; therefore majority of the contributions were reverted back. On the contrary, the evolution of high quality articles shows a significantly different pattern. In general, the number of contributions for high quality articles is larger compared to the number of contributions for low quality articles. It appears that in a particular period of time (2006 and 2007) the high quality articles became the focus of the Wikipedia community; thus the number of contributions rose with increasing maturity. Afterward, the articles became good or featured and the number of contributions decreased.

Generally, the number of reverted back modifications for high quality articles is smaller compared to the number of accepted contributions. However, after the articles became good or featured, the number of reverted back modifications exceeded the number of accepted contributions. This trend suggests that with increasing maturity, Wikipedia community tends to accept less number of new contributions. Therefore, a lot of contributions are reverted back when the articles are already of high quality.

6.3.2 Major and Minor Contributions

Wikipedia categorizes modifications as minor or major. A minor edit is one where the modification requires no review and could never be the subject of a dispute. An

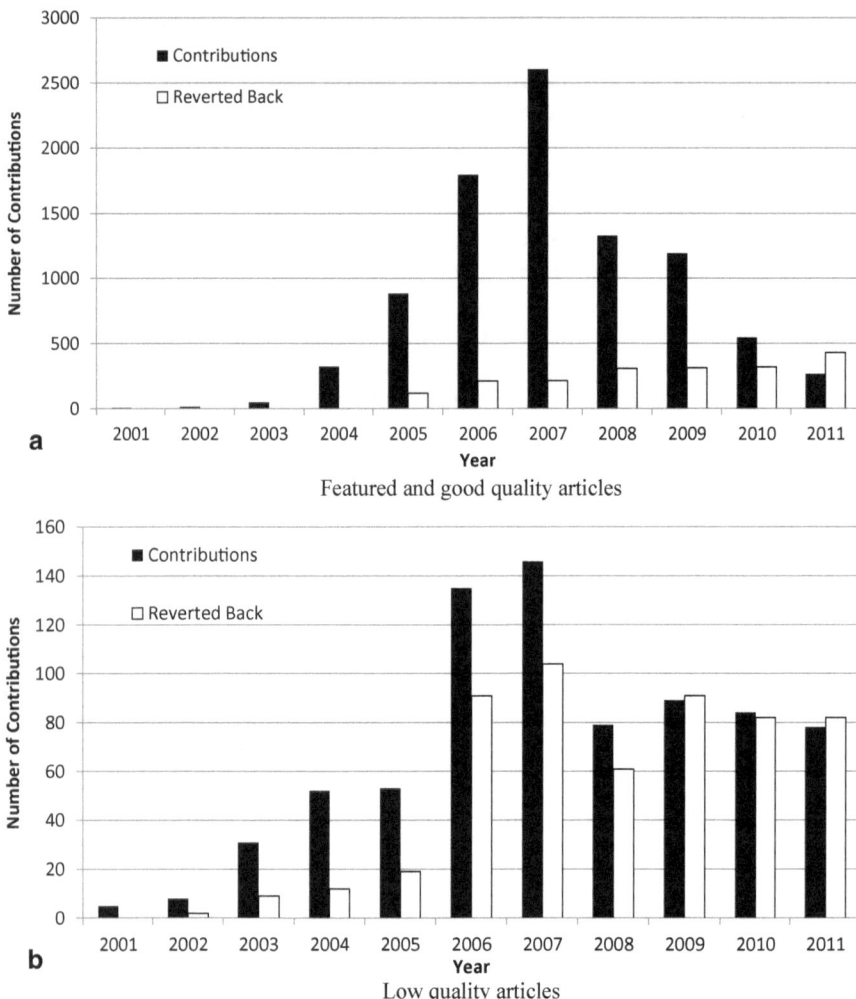

Fig. 6.1 Number of contributions for low and high quality articles. **a** Featured and good quality articles **b** Low quality articles

edit of this kind is marked in its page's revision history with a lower case, bold "m" character. However, a major edit is one that should be reviewed for its acceptability by all concerned editors. We could observe that most of the contributions for high quality articles are major (i.e., changing a big portion of an article). For high quality articles, around 85 % of contributions were tagged as major compared to only 52 % of the modifications being major for low quality articles as shown in Fig. 6.2.

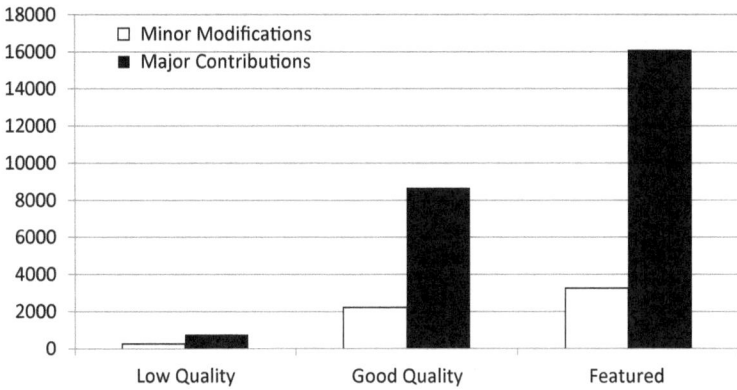

Fig. 6.2 Average number of major and minor contributions

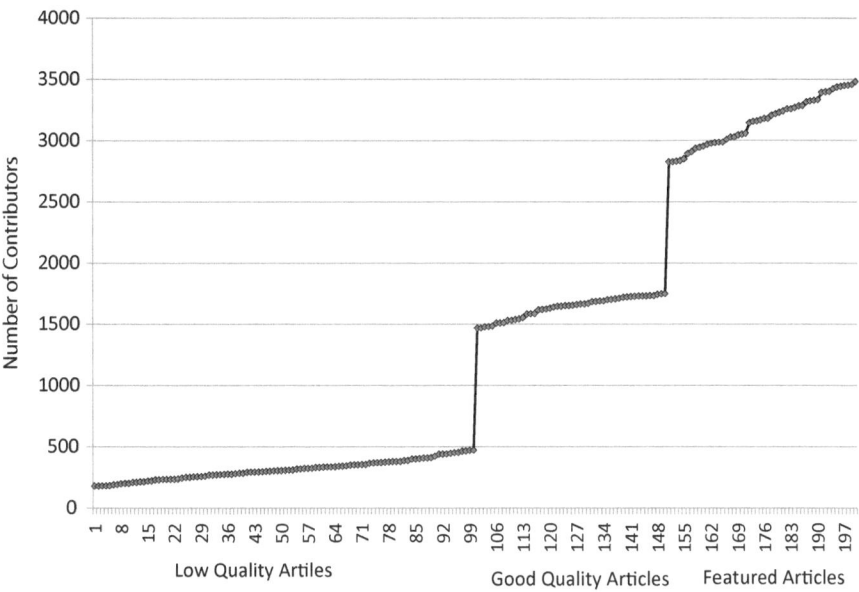

Fig. 6.3 Number of contributors for low quality, good quality, and featured articles

6.3.3   Contributors

We found the number of contributors for different types of documents as shown in Fig. 6.3. In this figure, we index the pages from 1 to 200 and sort them in terms of the number of contributors (horizontal label). The average number of contributors for low quality articles is 356 compared to 1,621 for good articles and 3,332 for featured ones. This shows that high quality articles gain wide exposure and receive

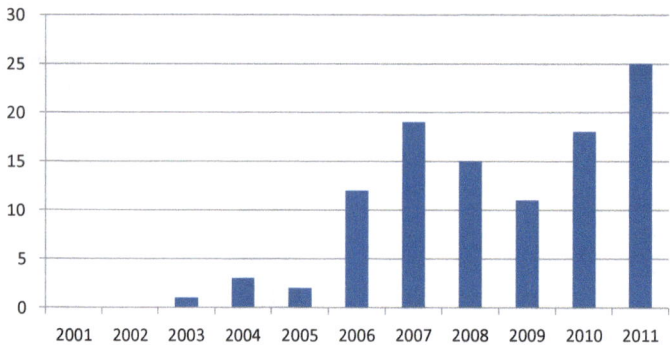

Fig. 6.4 Average number of reverting back done by top editors of high quality articles

about 10 times as many contributions as the low quality articles. In addition, we noticed that there is a highly active group of contributors involved from the creation of high quality articles until present. However, the majority of editors for low quality articles never contribute after their initial contributions.

We consider an editor as a top contributor for an article if the number of times her modifications got reverted back is much smaller than the total number of her contributions. In other words, the set of top contributors (E) for an article will be:

$$E = \{\forall u_i \in U \,|\, R_{u_i}/C_{u_i} < \epsilon, C_{u_i} > \Delta\}, \qquad (6.1)$$

where U is the set of all editors, R_{u_i} is the number of revert backs done on contributor u_i's modifications, C_{u_i} is the total number of contributions for editor u_i, and Δ and ϵ are the thresholds. We need the threshold in order not to consider editors who only contributed few times and all their modifications got accepted. We found out that for high quality articles the average number of top contributors is 32 for the threshold $\Delta = 50$ and $\epsilon = 0.1$. In other words, a very small group of contributors are responsible for the majority of activities around a high quality article. Henceforth, we focus on this small group of top contributors and analyze their activities and impacts.

6.3.4   Characterizing Top Contributors

To analyze the activities of top contributors in more detail, we measured the number and quality of their contributions (minor or major) as well as the number of revert backs. We noticed that they are responsible for more than 62 % of accepted contributions and around 85 % of revert back modifications for high quality articles. Fig. 6.4 shows the average number of revert backs done by top contributors. This figure shows an increase in the average number of revert backs through the years. In addition, we measured the quality of top contributors' modification by finding whether they were minor or major. We figured that more than 90 % of their contributions were tagged as

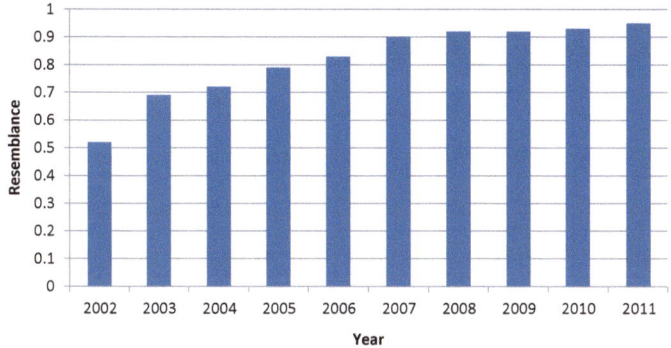

Fig. 6.5 Resemblance between top contributors

major compared to other contributors whose modifications were tagged minor more than 92 %. It seems that for high quality articles, top contributors formed informal groups which are responsible for the pages. Although Wikipedia is a democratic publishing platform which provides contributing capability for everyone, the top contributors tend to impose their opinions by taking "ownership" of the articles and reverting back others' modifications.

As another characteristic of top contributors, we examined the notion of resemblance to measure the similarity between two sets of top contributors in two consecutive years. We calculated the set of top contributors for each year separately in the interval 2001 to 2011. We define resemblance as the portion of top editors who remain in the set over two years. Let R_t denote the resemblance of two editing sets at year t. R_t can be defined as:

$$R_t = \left| \frac{E(t) \cap E(t+1)}{E(t)} \right|, \tag{6.2}$$

where $E(t)$ is the set of top contributors for a high quality article at year t. The value of R_t varies between 0 and 1 where $R_t = 1$ shows the case when the entire set of top contributors stayed for the next year, and $R_t = 0$ represents the case where none of the top contributors at year t remained in the subsequent year. Fig. 6.5 represents the average resemblance of top contributors for 100 high quality pages. It appears that the majority of top contributors were present from the creation of high quality articles until the present time. Also, after the documents reached their maturity and became featured or good articles in Wikipedia, the resemblance of top contributors in subsequent years is more than 90 %. In other words, the top contributors have become the owners of high quality articles and their engagement has increased.

Further, we investigated the similarity of top contributors in terms of their interests. For this purpose, we measured the overlap between the topics which have been mostly edited by top contributors. Table 6.1 presents the similarity of top contributors for high and low quality articles. Top contributors of high quality articles are more like-minded than the top contributors of low quality articles. We noticed that the similarity

Table 6.1 Similarity of top contributors

Contributors	Top 10	Top 20	Top 30	Top 40	Top 50
High quality articles	89.60 %	85.20 %	82.85 %	63.12 %	61.23 %
Low quality articles	65.02 %	63.34 %	42.24 %	29.86 %	25.01 %

Table 6.2 Similarity of top 10 contributors with bottom contributors

Editors	Bottom 10	Bottom 20	Bottom 30	Bottom 40	Bottom 50
High quality articles	42.51 %	40.89 %	33.23 %	28.23 %	33.44 %
Low quality articles	53.33 %	51.98 %	48.39 %	44.78 %	38.01 %

between top 10, 20, and 30 contributors of high quality articles is more than 80 %. This value suddenly drops to 60 % when the number of top contributors increases to 40. This is due to the fact that the top-30 contributors is close to the typical number of top contributors which amounts to 32 based on Eq. 6.1. Accordingly, Table 6.2 shows the similarity between top-10 and bottom-N ($N = 10, 20, ..., 50$) contributors for different types of articles. Here, bottom contributors are the ones who have the least number of contributions. We found out that the similarity between top and bottom contributors is very small. In addition, top editors for high quality articles are more focused on specific topics compared to the top editors for low quality articles. Therefore, the similarity between top editors and bottom editors for high quality articles is smaller compared to low quality articles.

6.4 Discussion

In this chapter, we analyzed the Wikipedia dataset for different types of articles based on the role of contributors. We found out that the main difference between low quality and featured articles is the number of contributions. High quality Wikipedia articles have more contributors than the lower quality ones. We noticed that for low quality articles, the number of contributions is very close to the number of revert backs modifications. Therefore, articles in this group suffer from instability and low integrity.

For high quality articles, we also observed that most of contributions are performed by a small number of contributors who have similar interests. They form a small group with size around 32 and edit similar topics in Wikipedia. In addition, this small groups of editors control the high quality documents and is responsible for reverting back the edits made by others. In other words, we can claim that a small set of users have taken ownership of the featured articles because of their contributions and tend to revert back other's contributions. This results in higher quality for a small portion of articles in Wikipedia. We observe that to have higher integrity in crowdsourcing systems, we need to have a permanent set of contributors who are dedicated for maintaining the quality of the contributions to the articles. For systems with open access such as Wikipedia, this can be a huge burden for the permanent editors. Therefore, we need new mechanisms for coordinating the activities in a crowdsourcing information system.

Chapter 7
Concluding Remarks

The tremendous success of crowdsourcing systems has attracted a lot of researchers. Prior work have investigated wide array of issues related to crowdsourcing such as visualization tools [177], [209], [210], motivations for participation [108], the effects of coordination and collaboration [217], vandalism analysis and detection [120], [160], [162], [185], reputation systems [27], [135], [223], quality assurance and automatic quality measurement [27], [52], [73], [82], [135], [193], [223].

For the purposes of this book, we interpret crowdsourcing quite broadly: any system functioning because of user contributed effort is a crowdsourcing system. With this broad system, we can see two major types of systems: content owned by the creators and content owned by the system. We discussed three major issues concerning crowdsourcing systems: identities, confidentiality control, and integrity management.

Identity is a key issue related to crowdsourcing. Without a well managed identity management scheme, a crowdsourcing system cannot keep mischief makers out of the system. If the mischief makers can dominate the activities, productive users will eventually leave the crowd, which will lead to the eventual collapse of the crowdsourcing system. We discussed three major types of identities: fixed identities, pseudonyms, and social identities. All three of them are suitable for crowdsourcing systems. The first two types are already in use in existing crowdsourcing systems. With the dominance of social networking in the Internet, it is becoming possible to use identities issued by OSN operators (e.g., Facebook and LinkedIn) to engage in crowd activities.

With creator-owned content, it becomes necessary to deal with confidentiality control in crowdsourcing systems. Online social networks are examples of crowd-sourcing systems that fall into this category of systems. In such a systems, users need mechanisms to manage the confidentiality constraints. Because the primary objective of crowdsourcing systems is not to manage the confidentiality of content generated in them, users need human friendly schemes that can automate most of the management chores. Ideally, this means setting the default policies to minimize the risk of unintended data leakages and also alert the users if there are any anomalous data sharing patterns.

A. Ranj Bar, M. Maheswaran, *Confidentiality and Integrity in Crowdsourcing Systems*, SpringerBriefs in Applied Sciences and Technology, DOI 10.1007/978-3-319-02717-3_7, © The Author(s) 2014

The other important data protection issue is integrity management. Integrity management is a important issue in crowdsourcing systems where the content is system owned. With system-owned content, contributions are solicited from the crowd to evolve the state of the content. To regulate the quality of the content, integrity management is management is essential. We discussed two different types of approaches for integrity management: content-centric approaches and contributor-centric approaches.

References

1. Reliability of Wikipedia, http://en.wikipedia.org/wiki/Reliability_of_Wikipedia
2. Http://en.wikipedia.org/wiki/Special:Statistics
3. The sheep market. http://www.thesheepmarket.com/
4. Amazon mechanical turk. https://www.mturk.com/
5. Online social networks—research report, http://www.communities.gov.uk/publications/communities/onlinesocialnetworks, October 2008
6. After 10 Years of Blogs, the Future's Brighter Than Ever, http://www.wired.com/entertainment/theweb/news/2007/12/blog-anniversary
7. Alexa Top 500 Global Sites, http://www.alexa.com/topsites
8. For Social Networks, There's Still Room to Play, http://www.nielsen-online.com/blog/2008/10/22/for-social-networks-there's-still-room-to-play/
9. Http://en.wikipedia.org/wiki/Wikipedia:Featuredarticles
10. Http://en.wikipedia.org/wiki/User:ClueBotNG
11. Http://en.wikipedia.org/wiki/User:XLinkBot
12. Advogatos trust metric. in http://www.advogato.org/trust-metric.html.
13. Blogger. http://www.blogger.com/start.
14. Citizendium, http://en.citizendium.org/
15. http://en.wikipedia.org/wiki/real_id_act
16. http://en.wikipedia.org/wiki/social_identity
17. https://cms.paypal.com/us/cgi-bin/?&cmd=_render-content&content_id=ua/useragreement_full
18. http://www.neobux.com/
19. http://www.neobux.com/m/a/
20. Scholarpedia, http://www.scholarpedia.org/
21. The security limitations of SSO in Openid, author=H. Oh and S. Jin, booktitle=10th IEEE International Conference on Advanced Communication Technology, volume=3, pages=1608–1611, year=2008,
22. The atom syndication format. IETF Internet Draft (awaiting Internet Eng. Steering Group evaluation) (2005)
23. http://www.nist.gov/nstic/nstic-why-we-need-it.pdf (2011)
24. Abdul-Rahman A, Hailes S (1997) Using recommendations for managing trust in distributed systems. In: IEEE Malaysia International Conference on Communication
25. Abdul-Rahman A, Hailes S (2000) Supporting trust in virtual communities. In: 3rd Hawaii International Conference on System Sciences
26. Adar E, Huberman B (2000) Free riding on gnutella. First Monday 5(10):2–13
27. Adler BT, de Alfaro L (2007) A content-driven reputation system for the wikipedia. In: 16th International Conference on World Wide Web, pp 261–270
28. Adler BT, Chatterjee K, de Alfaro L, Faella M, Pye I, Raman V (2008) Assigning trust to wikipedia content. In: 2008 International Symposium on Wikis

A. Ranj Bar, M. Maheswaran, *Confidentiality and Integrity in Crowdsourcing Systems,*   69
SpringerBriefs in Applied Sciences and Technology, DOI 10.1007/978-3-319-02717-3,
© The Author(s) 2014

29. Agarwal N, Liu H, Murthy S, Sen A, Wang X (2009) A social identity approach to identify familiar strangers in a social network. In: 3rd International AAAI Conference of Weblogs and Social Media

30. Aggarwal C, Wolf JL, Wu KL, Yu PS (1999) Horting hatches an egg: a new graph-theoretic approach to collaborative filtering. In: 5th ACM SIGKDD Conference on Knowledge Discovery and Data Mining

31. Ahmed N, Hadaller D, Keshav S (2005) An incremental approach for maintaining upto-date global aggregates. Technical report, University of Waterloo

32. von Ahn L (2006) Games with a purpose. IEEE Computer 39(6):92–94

33. von Ahn L, Dabbish L (2004) Labeling images with a computer game. In: International Conference on Human Factors in Computing Systems. Vienna

34. von Ahn L, Liu R, Blum M (2006) Peekaboom: a game for locating objects in images. In: ACM SIGCHI Conference on Human Factors in Computing Systems

35. von Ahn L, Maurer B, McMillen C, Abraham D, Blum M (2008) Recaptcha: human-based character recognition via web security measures. Science 321(5895):1465–1468

36. Akcora C, Carminati B, Ferrari E (2012) Privacy in social networks: how risky is your social graph? In: 28th International Conference on Data Engineering

37. Akcora C, Carminati B, Ferrari E (2012) Risks of friendships on social networks. In: IEEE International Conference on Data Mining

38. Akerlof G, Kranton R (2000) Economics and identity. Q J Econ 115(3):715–753

39. Ali B, Villegas W, Maheswaran M (2007) A trust based approach for protecting user data in social networks. In: Conference of the IBM Center for Advanced Studies on Collaborative Research, pp 288–293

40. Andrade N, Mowbray M, Lima A, Wagner G, Ripeanu M (2005) Influences on cooperation in bittorrent communities. In: ACM SIGCOMM Workshop on Economics of Peer-To-Peer Systems

41. Anwar M, Fong PWL (2012) A visualization tool for evaluating access control policies in facebook-style social network systems. In: 27th ACM Symposium on Applied Computing (SAC'12), Security Track

42. Anwar M, Fong PWL, Yang XD, Hamilton H (2009) Visualizing privacy implications of access control policies in social network systems. In: 4th International Workshop on Data Privacy Management, pp 106–120

43. Bell D, Lapadula L: Secure computer systems: mathematical foundations. Technical Report MTR-2547, The MITRE Corporation

44. Bell D, Lapadula L (1976) Secure computer system: unified exposition and multics interpretation. Deputy for Command and Management Systems, United State Air Force

45. Benantar M (2006) Access Control Systems: Security, Identity Management and Trust Models, Springer, New York

46. Berkman LF, Syme SL (1979) Social networks, host resistance, and mortality: a nine-year follow-up study of alameda county residents. Am J Epidemiol 109(2):186–204

47. Bilge L, Strufe T, Balzarotti D, Kirda E (2009) All your contacts belong to us: automated identity theft attacks on social networks. In: 18th ACM International Conference on World wide web, pp 551–560

48. Binzel C, Fehr D (2010) Social Relationships and Trust. Discussion Papers of DIW Berlin

49. Bishop M (2002) Computer Security: Art and Science. Addison-Wesley

50. Blaze M, Feigenbaum J (1999) Managing trust in information labeling system. European Transactions on Telecommunications 8:491–501

51. Blaze M, Feigenbaum J, Lacy J (1996) Decentralized trust management. In: IEEE Conference on Security and privacy

52. Blumenstock JE (2008) Size matters: word count as a measure of quality on wikipedia. In: 17th International Conference on World Wide Web, pp 1095–1096

53. Bonneau J, Anderson J, Anderson R, Stajano F (2009) Eight friends are enough: social graph approximation via public listings. In: 2nd ACM EuroSys Workshop on Social Network Systems, pp 13–18

54. Boyd DM, Ellison NB (2008) Social network sites: definition, history, and scholarship. J Comput Mediat Commun 13(1):210–230
55. Brabham DC (2008) Crowdsourcing as a model for problem solving: an introduction and cases. Convergence: The International Journal of Research into New Media Technologies 14(1):75–90
56. Brainard J, Juels A, Rivest R, Szydlo M, Yung M (2006) Fourth-factor authentication: somebody you know. In: 13th ACM Conference on Computer and Communications Security, pp 168–178
57. Bruns G, Fong PWL, Siahaa I, Huth M (2012) Relationship-based access control: its expression and enforcement through hybrid logic. In: 2nd ACM Conference on Data and Application Security and Privacy, pp 51–60
58. Buchegger S, Boudec JYL (2003) The effect of rumor spreading in reputation systems for mobile ad-hoc networks. In: Workshop on Modeling and Optimization in Mobile, Ad Hoc and Wireless Networks
59. Bullock A, Benford S (1999) An access control framework for multi-user collaborative environments. In: ACM SIGGROUP Conference on Supporting Group Work, pp 140–149
60. Buskens V (2002) Social Networks and Trust. Kluwer academic Publishers, Norwell
61. Cachia R, Compano R, Costa OD (2007) Technological forecasting and social change. Communications of the ACM 74(8):1179–1203
62. Carminati B, Ferrari E, Girardi J (2012) Trust and share: Trusted information sharing in online social networks. In: 28th International Conference on Data Engineering
63. Carminati B, Ferrari E, Morasca S, Taibi D (2011) A probability-based approach to modeling the risk of unauthorized propagation of information in on-line social networks. In: 1st ACM Conference on Data and Application Security and Privacy, CODASPY '11, pp 51–62
64. Carminati B, Ferrari E, Perego A (2006) Rule-based access control for social networks. In: OTM Workshops, pp 1734–1744
65. Carminati B, Ferrari E, Perego A (2009) Enforcing access control in web-based social networks. ACM Transactions on Information and System Security 13:6:1–6:38
66. Chang Y, Chakrabarti D, Wang C, Falautsos C (2003) Epidemic spreading in real networks: an eigenvalue viewpoint. In: 22nd International Symposium on Reliable Distributed Systems
67. Chaum D (1985) Security without identification: transaction systems to make big brother obsolete. Communications of the ACM 28(10):1030–1044
68. Chen G, Rahman F (2008) Analyzing privacy designs of mobile social networking applications. In: IEEE/IFIP International Conference on Embedded and Ubiquitous Computing, vol. 2, pp 83–88
69. Chen Z, Gao L, Kwiat K (2003) Modeling the spread of active worms. In: IEEE International Conference on Computer Communications
70. Cheriton D (1992) Dissemination-oriented communication systems. Technical report, Stanford University
71. Chiang M (2012) Networked Life: 20 Questions and Answers. Cambridge University Press, Cambridge
72. Choi H, Kruk S, Grzonkowski S, Stankiewicz K, Davis B, Breslin J (2006) Trust models for community-aware identity management. In: Identity, Reference, and the Web Workshop
73. Cross T (2006) Puppy smoothies: Improving the reliability of open, collaborative wikis. First Monday 11(9)
74. Damiani E, Vimercati SDC, Paraboschi S, Samarati P, Violante F (2002) A reputation-based approach for choosing reliable resources in peer-to-peer networks. In: 9th ACM Conference on Computer and Communications Security
75. De AP, Schorlemmer M, Csic I, Cranefield S (2010) A Social-Network Defence against Whitewashing. In: 9th International Conference on Autonomous Agents and Multiagent Systems, pp 1563–1564
76. Demers A, Greene D, Hauser C, Irish W, Larson J, Shenker S, Sturgis H, Swinehart D, Terry D (1987) Epidemic algorithms for replicated database maintenance. In: 6th ACM Symposium on Principles of Distributed Computing

77. DeRose P, Chai X, Gao BJ, Shen W, amd P, Bohannon AD, Zhu X (2008) Building community wikipedias: a machine-human partnership approach. In: 24th IEEE International Conference on Data Engineering

78. Deutsh M (1962) Cooperation and trust: Some theoritical notes. In: MR Jones (ed.) Nebraska Symposium on Motivation,Nebraska University Press, pp 275–320

79. Dingledine R, Mathewson N, Syverson P (2003) Reputation in p2p anonymity systems. In: Workshop on Economics of P2P Systems

80. Doan A, Ramakrishnan R, Halevy AY (2011) Crowdsourcing systems on the world-wide web. Communications of the ACM, 54(4):86–96

81. Domingo-Ferrer J, Viejo A, Sebé F, González-Nicolás U (2008) Privacy homomorphisms for social networks with private relationships. Computer Networks: The International Journal of Computer and Telecommunications Networking 52:3007–3016

82. Dondio P, Barrett S (2007) Computational trust in web content quality: a comparative evalutation on the wikipedia project. Informatica–An International Journal of Computing and Informatics 31(2):151–160

83. Douceur J (2002) The sybil attack. Peer-to-Peer Systems, pp 251–260

84. Eaton B, Eswaran M, Oxoby R: us and them: (2009) The origin of identity, and its economic implications. Working Papers

85. Elahi N, Chowdhury MMR, Noll J (2008) Semantic access control in web based communities. In: 3rd International Multi-Conference on Computing in the Global Information Technology, pp 131–136

86. Eugster PT, Guerraoui R, Kermarrec AM, Massouli L (2004) Epidemic information dissemination in distributed systems. IEEE Computer 37(5):60–67

87. Ferraiolo D, Kuhn R (1992) Role-based access controls. In: 15th NIST-NCSC National Computer Security Conference, pp 554–563

88. Fong PW (2011) Relationship-based access control: protection model and policy language. In: 1st ACM Conference on Data and Application Security and Privacy, pp 191–202

89. Fong PWL, Anwar M, Zhao Z (2009) A privacy preservation model for facebook-style social network systems. In: 14th European Conference on Research in Computer Security, pp 303–320

90. Fong PWL, Siahaa I (2011) Relationship-based access control policies and their policy languages. In: 16th ACM Symposium on Access Control Models and Technologies, pp 51–60

91. Ford B, Strauss J (2008) An offline foundation for online accountable pseudonyms. 1st ACM Workshop on Social Network Systems, pp 31–36

92. Friedman E, Resnick P (2000) The social cost of cheap pseudonyms. J Econ Manag Strategy 10:173–199

93. Fu K, Kaashoek MF, Mazieres D (2002) Fast and secure distributed read-only file system. ACM Transactions on Computer Systems 20(1):1–24

94. Gambetta D (2000) We trust trust? Technical report, Department of Sociology, University of Oxford

95. Ganesh A, Kermarrec A, Massoulie L (2003) Peer-to-peer membership management for gossip-based protocols. IEEE Transactions on Computers 52(2):139–149

96. Garetto M, Gong W, Towsley D (2003) Modeling malware spreading dynamics. In: IEEE International Conference on Computer Communications

97. Gkantsidis C, Mihail M, Saberi A (2004) Random walks in peer-to-peer networks. In: IEEE International Conference on Computer Communications

98. Golbeck J, Hendler J (2006) Inferring binary trust relationships in web-based social networks. ACM Transactions on Internet Technology 6(4):529

99. Goldenberg J, Libai B, Muller E (2001) Talk of the network A complex systems look at the underlying process of word-of-mouth. Marketing Letters 3(12):211–223

100. Grandison T (2003) Trust management for internet applications. Ph.D. thesis, Imperial College London

101. Grandison T, Sloman M (2000) A survey of trust in internet applications. IEEE Communications Surveys

102. Granovetter M (1978) Threshold models of collective behavior. Am J Sociology 83(6): 1420–1443
103. Granovetter MS (1973) The strength of weak ties. American Journal of Sociology 78(6): 1360–1380
104. Habib A, Chuang J (2004) Incentive mechanism for peer-to-peer media streaming. In: 12th IEEE International Workshop on Quality of Service
105. Hammond T, Hannay T, Lund B (2004) The role of rss in science publishing. D-Lib Magazine 10(12)
106. Hart M, Johnson R, Stent A (2007) More content—less control: Access control in the web 2.0. In: Web 2.0 Security and Privacy Workshop
107. He J, Chu W, Liu Z (2006) Inferring privacy information from social networks. Intelligence and Security Informatics, pp 154–165
108. Hoisl B, Aigner W, Miksch S (2007) Social rewarding in wiki systems—motivating the community. In: 2nd Online Communities and Social Computing, pp 362–371
109. Ives ZG., Khandelwal N, Kapur A, Cakir M (2005) Orchestra: rapid, collaborative sharing of dynamic data. In: Conference on Innovative Data Systems Research
110. Jennings B, Finkelstein A (2009) Digital identity and reputation in the context of a bounded social ecosystem. In: Business Process Management Workshops, pp 687–697x
111. Jin L, Takabi H, Joshi J (2011) Towards active detection of identity clone attacks on online social networks. In: 1st ACM Conference on Data and Application Security and Privacy, pp 27–38
112. Josang A, Ismail R, Boyd C (2005) A survey of trust and reputation systems for online service provision. Decision Support Systems
113. Jurvetson S, Draper R: Viral marketing. Netscape M-Files (June 1997)
114. Kamvar SD, Schlosser MT, Garcia-Molina H (2003) The eigentrust algorithm for reputation management in P2Pnetworks. In: 12th International World Wide Web Conference
115. Kang MH, Park JS, Froscher JN. (2001) Access control mechanisms for inter-organizational workflow. In: 6th ACM Symposium on Access Control Models and Technologies, pp 66–74
116. Kasneci G, Ramanath M, Suchanek M, Weiku G (2008) The yago-naga approach to knowledge discovery. SIGMOD Rec 37(4):41–47
117. Kempe D, Dobra A, Gehrke J (2003) Gossip-based computation of aggregate information. In: 44th Annual IEEE Symposium on Foundations of Computer Science
118. Kephart JO, White SR (1993) Measuring and modeling computer virus prevelance. In: IEEE Computer Society Symposium on Research In Security and Privacy, pp 2–15
119. Kimball L, Rheingold H (2000) How online social networks benefit organizations. Rheingold Associates
120. Kittur A, Suh B, Pendleton BA, Chi EH (2007) He says, she says: conflict and coordination in wikipedia. In: SIGCHI Conference on Human Factors in Computing Systems, pp 453–462
121. Koblin AM (2009) The sheep market. In: Seventh ACM Conference on Creativity and Cognition, pp 451–452
122. Konstan J, Miller B, Maltz D, Herlocker J (1997) Grouplens Applying collaborative filtering to usenet news. Communications of ACM 40(3):77–87
123. Konstan J, Miller B, Maltz D, Herlocker J (2005) Emerging technologies: Blogs and wikis: Environments for on-line collaboration. Language Learning & Technology 7(2):12–16
124. Krasnova H, Gunther O, Spiekermann S, Koroleva K (2009) Privacy concerns and identity in online social networks. Identity in the Information Society 2(1):39–63
125. Kruk SR, Grzonkowski S, Gzella A, Woroniecki T, Choi H (2006) D-foaf: distributed identity management with access rights delegation. In: Asian Semantic Web Conference, Lecture Notes in Computer Science, vol. 4185, Springer, pp 140–154
126. Lampe C, Resnick P (2004) Slash(dot) and burn: distributed moderation in a large online conversation space. In: ACM SIGCHI Conference on Human Factors in Computing Systems
127. Lampson BW (1971) Protection. In: 5th Princeton Symp. Information Science and Systems, pp 437–443

128. Law E, von Ahn L (2009) Input-agreement: A new mechanism for data collection using human computation games. In: International Conference on Human Factors in Computing Systems
129. Lee S, Sherwood R, Bhattacharjee B (2003) Cooperative peer groups in NICE. In: IEEE International Conference on Computer Communications
130. Leimeister J, Huber M, Bretschneider U, Krcmar H (2009) Leveraging crowdsourcing: activation-supporting components for it-based ideas competition. Journal of Management Information Systems 26:197–224
131. Leskovec J, Adamic LA (2007) The dynamics of viral marketing. ACM Transactions on the Web 1(1)
132. Leskovec J, Huttenlocher D, Kleinberg J (2010) Predicting positive and negative links in online social networks, pp 641–650
133. Levine B, Shields C, Margolin N (2006) A survey of solutions to the sybil attack. University of Massachusetts Amherst
134. Lih A (2004) Wikipedia as participatory journalism: Reliable sources? metrics for evaluating collaborative media as a news resource. In: 5th International Symposium on Online Journalism
135. Lim EP, Vuong BQ, Lauw HW, Sun A (2006) Measuring qualities of articles contributed by online communities. In: IEEE/WIC/ACM International Conference on Web Intelligence, pp 81–87
136. Lysyanskaya A, Rivest R, Sahai A, Wolf S (2000) Pseudonym systems. Selected Areas in Cryptography, Springer, pp 184–199
137. Maheswaran M, Ali B, Ozguven H, Lord J: Online identities and social networking. In: B. Furht (ed.) Handbook of Social Network Technologies and Applications
138. Mandel M, Ellis D (2007) A web-based game for collecting music metadata. In: 8th International Conference on Music Information Retrieval
139. Mankoff J, Matthews D, Fussell S, Johnson M (2007) Leveraging social networks to motivate individuals to reduce their ecological footprints. HICSS 07 Proceedings of the 40th Annual IEEE Hawaii International Conference on System Sciences, p 87
140. Marsh S (1994) Formalizing trust as a computer concept. Phd thesis, Department of Computer Science and Mathematics, University of Stirling
141. Marshall A, Tompsett B (2005) Identity theft in an online world. Computer Law and Security Report 21(2):128–137
142. McCann R, Doan A, Varadarajan V, Kramnik A (2003) Building data integration systems: A mass collaboration approach. In: International Workshop on Web and Databases
143. McCann R, Shen W, Doan A (2008) Matching schemas in online communities: a web 2.0 approach. In: 24th IEEE International Conference on Data Engineering
144. Menon A (2011) Free Facebook t-shirts at the cost of your Personal Information? Security Advisor Research Blog, http://totaldefense.com/securityblog/2011/09/09/Free-Facebook-t-shirts-at-the-cost-of-your-Personal-Information.aspx
145. Mislove A, Marcon M, Gummadi KP, Druschel P, Bhattacharjee B (2007) Measurement and analysis of online social networks. In: 5th ACM/USENIX Internet Measurement Conference (IMC'07)
146. Mitra A, Maheswaran M (2007) Impact of peer churning on trusted gossiping for peer-to-peer information sharing. In: Workshop on Trust and Reputation Management in Massively Distributed Computing Systems (TRAM), (held in conjunction with IEEE ICDCS), Toronto, pp 702–707
147. Mitra A, Maheswaran M (2007) Trusted gossip: A rumor resistance dissemination mechanism for peer-to-peer information sharing. In: 21st IEEE International Conference on Advanced Information Networking and Applications, Niagara Falls, Canada, pp 702–707
148. Mitra A, Udupa R, Maheswaran M (2005) A secured hierarchical trust management framework for public computing utilities. In: Conference of the IBM Center for Advanced Studies on Collaborative Research
149. Montgomery JD.: Social networks and labor-market outcomes: Toward an economic analysis. The American Economic Review 81(5), 1408–1418
150. Montresor A, Jelasity M, Babaoglu O (1991) Robust aggregation protocols for largescale overlay networks. In: International Conference on Dependable Systems and Networks (DSN) (June 2004)

151. Moore D, Shannon C, Voelker GM, Savage S (2003) Internet quarantine: Requirements for containing self-propagating code. In: IEEE International Conference on Computer Communications
152. Morris M (1993) Epidemiology and social networks: modeling structured diffusion. Sociological Methods Research 22(1):99–126
153. Neubauer T, Riedl B (2008) Improving patients privacy with pseudonymization. Studies in health technology and informatics 136:691
154. Olson M (2008) The amateur search. SIGMOD Rec 37(2):21–24
155. Ong J (2011) UK tribunal upholds Apple's firing of retail employee for critical Facebook post. Apple Insider, http://www.appleinsider.com/articles/11/11/01/uktribunalupholdsapples firingofretailemployeeforcriticalfacebookpost.html
156. Pallickara SL., Plale B, Fang L, Gannon D (2006) Trust cell Towards the end-to-end trust in data-oriented scientific computing. In: Parallel Processing Workshops
157. Park J, Sandhu R (2004) The $UCON_{ABC}$ usage control model. ACM Transactions on Information and System Security 7:128–174
158. Pashalidis A, Mitchell C (2006) Limits to anonymity when using credentials. Security Protocols, pp 13–19
159. Penna P, Schoppmann F, Silvestri R, P.Widmayer (2009) Pseudonyms in cost-sharing games. Internet and Network Economics, pp 256–267
160. Potthast M, Stein B, Gerling R (2008) Automatic vandalism detection in wikipedia. In: Advances in Information Retrieval—30th European Conference on IR Research, pp 663–668
161. Praetorius D (2011) Facebook "Phonebook Contacts" Stores Your Friends' Phone Numbers But Doesn't Share Them. The Huffington Post, Tech Canada, http://www.huffingtonpost. com/2011/08/11/facebook-phonebook-contactsn924543.html.html
162. Priedhorsky R, Chen J, Lam STK, Panciera K, Terveen L, Riedl J (2007) Creating, destroying, and restoring value in wikipedia. In: 2007 International ACM Conference on Supporting Group Work
163. Pujol J, Sanguessa R (2002) Extracting reputation in multi-agent systems by means of social network topology. In: 1st International Conference on Autonomous Agents and Multi-Agent Systems
164. Ranjbar A, Maheswaran M (2010) A case for community-centric controls for information sharing on online social networks. In: IEEE GLOBECOM Workshop on Complex and Communication Networks
165. Ranjbar A, Maheswaran M (2011) Blocking in community-centric information management approaches for the social web. In: IEEE Global Communications Conference
166. Ranjbar A, Maheswaran M (2011) Community-centric approaches for confidentiality management in online systems. In: 20th IEEE International Conference Computer Communication Networks
167. Recordon D, Reed D (2006) Openid 2.0: a platform for user-centric identity management. In: 2nd ACM Workshop on Digital Identity Management, pp 11–16
168. Resnick P, Varian HR (1997) Recommender systems. Communications of ACM 40(3):56–58
169. Resnick P, Zechhauser R (2000) Reputation system. Communications of ACM 43(12): 45–48
170. Resnick P, Zeckhauser R (2002) Trust among strangers in internet transactions: Empirical analysis of ebays reputation system. Advances in Applied Microeconomics
171. Rheingold H (2003) Smart Mobs. Perseus Publishing
172. Richardson M, Domingos P (2003) Building large knowledge bases by mass collaboration. In: International Conference on Knowledge Capture
173. Ripeanu M (2001) Peer-to-peer architecture case study: Gnutella network. In: 1st International Conference on Peer-to-Peer Computing
174. Risson J, Moors T (2004) Survey of research towards robust peer-to-peer networks: Search methods. Technical Report UNSW-EE-P2P-1-1, University of New South Wales
175. Rowstron A, Druschel P (2001) Pastry: Scalable, decentralized object location, and routing for large-scale peer-to-peer systems. In: IFIP/ACM International Conference on Distributed Systems Platforms (Middleware), pp 329–350

176. Sabater J, Sierra C (2002) Regret A reputation model for gregarious socities. In: 4th Workshop on Deception, Fraud and Trust in Agent Societies
177. Sabel M (2007) Structuring wiki revision history. In: 2007 International Symposium on Wikis, pp 125–130
178. Salz R (1992) Internetnews: Usenet transport for internet sites. USENIX
179. Sandhu RS (1993) Lattice-based access control models. IEEE Computer 26(11):9–19
180. Sandhu RS, Coyne EJ, Feinstein HL, Youman CE (1996) Role-based access control models. IEEE Computer 29(2):38–47
181. Sandhu RS, Samarati P (1994) Access control: Principles and practice. IEEE Communications Magazine 32(9):40–48
182. Saroiu S, Gummadi KP, Dunn R, Gribble SD, Levy HM (2002) An analysis of internet content delivery systems. In: 5th Symposium on Operating Systems Design and Implementation
183. Sarwar BM, Karypis G, Konstan JA, Riedl J (2001) Item-based collaborative filtering recommendation algorithms. In: 10th International World Wide Web Conference
184. Sit E, Morris R (2002) Security considerations for peer-to-peer distributed hash tables. In: 1st International Peer to Peer Systems Workshop
185. Smets K, Goethals B, Verdonk B (2008) Automatic vandalism detection in wikipedia: towards a machine learning approach. In: AAAI Workshop, Wikipedia and Artificial Intelligence: an Evolving Synergy
186. Soleymani B, Maheswaran M (2009) Social authentication protocol for mobile phones. In: IEEE International Conference on Computational Science and Engineering, vol. 4, pp 436–441
187. Squicciarini AC, Shehab M, Wede J (2010) Privacy policies for shared content in social network sites. VLDB J 19:777–796
188. Staab S, Bhargava BK, Lilien L, Rosenthal A, Winslett M, Sloman M, Dillon T, Chang FKHE, Nejdl W, Olmedilla D, Kashyap V (2004) The pudding of trust. Communications of ACM 19(5):74–88
189. Steffen A (2012) The linux integrity measurement architecture and tpm-based network endpoint assessment. Technical report, HSR University of Applied Sciences
190. Steinmetz R, Wehrle K (2005) Peer-to-peer systems and applications. Lecture Notes in Computer Science, Springer, pp 34–85
191. Stoica I, Morris R, Karger D, Kaashoek F, Balakrishnan H (2001) Chord: a scalable peer-to-peer lookup service for internet applications. In: 2001 ACM SIGCOMM Conference, pp 149–160
192. Stutzman F (2006) An evaluation of identity-sharing behavior in social network communities. International Digital and Media Arts Journal 3(1):10–18
193. Stvilia B, Twidale MB, Smith LC, Gasser L (2005) Assessing information quality of a community-based encyclopedia. In: International Conference on Information Quality, pp 442–454
194. Subramani M, Rajagopalan B (2003) Knowledge-sharing and influence in online social networks via viral marketing. Communications of the ACM 46(12):300–307
195. Surowiecki J (2005) The Wisdom of Crowds. Anchor Books
196. Suryanarayana G, Taylor RN (2004) A survey of trust management and resource discovery technologies in peer-to-peer applications. Isr technical report, Institute for Software Research
197. Suzuki Y, Yoshikawa M (2012) Mutual evaluation of editors and texts for assessing quality of wikipedia articles. In: 8th International Symposium on Wikis and Open Collaboration
198. Sweeney L et al (2002) k-anonymity: A model for protecting privacy. International Journal of Uncertainty Fuzziness and Knowledge Based Systems, 10(5):557–570
199. Tapscott D, Williams AD (2006) Wikinomics. Portfolio
200. Thomas RK (1997) Team-based access control: a primitive for applying role-based access controls in collaborative environments. In: 2nd ACM Workshop on Role-based Access Control, pp 13–19
201. Thomas RK, Sandhu RS (1998) Task-based authorization controls (tbac): A family of models for active and enterprise-oriented authorization management. In: IFIP TC11 WG11.3 Eleventh International Conference on Database Securty XI, pp 166–181

202. Tolone W, Ahn GJ, Pai T, Hong SP (2005) Access control in collaborative systems. ACM Computing Survey, 37(1):29–41
203. Tootoonchian A, Gollu KK, Saroiu S, Ganjali Y, Wolman A (2008) Lockr: social access control for web 2.0. In: 1st Workshop on Online Social Networks, pp 43–48
204. Tootoonchian A, Saroiu S, Wolman A, Ganjali Y (2009) Lockr Better privacy for social networks. In: 5th International Conference on Emerging Networking Experiments and Technologies
205. Travers J, Milgram S (1969) An experimental study of the small world problem. Journal of Sociometry 32:425–443
206. Turnbull D, Liu R, Barrington L, Lanckriet G (2007) A gamebased approach for collecting semantic annotations of music. In: 8th International Conference on Music Information Retrieval
207. Villegas W, Ali B, Maheswaran M (2008) An access control scheme for protecting personal data. In: 6th Annual Conference on Privacy, Security and Trust, pp 24–35
208. Viswanath B, Post A, Gummadi K, Mislove A (2010) An analysis of social network-based sybil defenses. ACM SIGCOMM Computer Communication Review 40(4):363–374
209. Vegas, F, Wattenberg M, Dave K (2004) Studying cooperation and conflict between authors with history flow visualizations. In: SIGCHI Conference on Human Factors in Computing Systems, pp 575–582
210. Vegas, F, Wattenberg M, Kriss J, Ham F (2007) Talk before you type: coordination in wikipedia. In: 40th Hawaii International Conference on System Sciences, pp 78–88
211. Wasserman S, Faust K (1994) Social Network Analysis: methods and Applications. Cambridge University Press, Cambridge
212. Watts D, Dodds P, Newman M (2002) Identity and search in social networks. Science 296(5571):1302–1305
213. Watts DJ, Strogatz SH (1998) Collective dynamics of small-world networks. Nature 393(6684):440–442
214. Weld DS, Wu F, Adar E, Amershi S, Fogarty J, Hoffmann R, Patel K, Skinner M (2008) Intelligence in wikipedia. In: 23rd AAAI Conference on Artificial Intelligence. Chicago
215. Wellman B, Berkowitz SD (1988) Social Structures: a Network Approach. Cambridge University Press, Cambridge
216. West A, Lee I (2012) Towards content-driven reputation for collaborative code repositories. In: 8th International Symposium on Wikis and Open Collaboration
217. Wilkinson DM, Huberman BA (2007) Cooperation and quality in wikipedia. In: 2007 International Symposium on Wikis, pp 157–164
218. Xu Z, Zhang Z (2002) Building low-maintenance expressways for p2p systems. Technical Report HPL-2002–41, HP Laboratories Palo Alto
219. Yokoo M, Sakurai Y, Matsubara S (2004) The effect of false-name bids in combinatorial auctions: new fraud in internet auctions. Games and Economic Behavior 46(1):174–188
220. Yu H, Gibbons PB, Kaminsky M, Xiao F (2008) Sybillimit: a near-optimal social network defense against sybil attacks. In: IEEE Symposium on Security and Privacy, pp 3–17
221. Yu H, Kaminsky M, Gibbons P, Flaxman A (2006) Sybilguard: defending against sybil attacks via social networks. ACM SIGCOMM Computer Communication Review 36(4):267–278
222. Yu H, Kaminsky M, Gibbons PB, Flaxman A (2006) Sybilguard: defending against sybil attacks via social networks. In: 2006 Conference on Applications, Technologies, Architectures, and Protocols for Computer Communications, pp 267–278. ACM
223. Zeng H, Alhossaini MA, Ding L, Fikes R, McGuinness DL (2006) Computing trust from revision history. In: International Conference on Privacy, Security and Trust: Bridge the Gap Between PST Technologies and Business Services, pp 8:1–8:1
224. Zhang D, Prior K, Levene M (2012) How long do wikipedia editors keep active? In: 8th International Symposium on Wikis and Open Collaboration
225. Zhang X, Nakae M, Covington MJ, Sandhu R (2008) Toward a usage-based security framework for collaborative computing systems. ACM Transactions on Information and System Security 11